COLOR TILE MATH

Grades K-3

Olga Gonzalez-Granat

ISBN: 1-56911-979-1

Printed in the United States of America.

TABLE OF CONTENTS

INTRODUCTION

Color Tile Math helps children explore basic math concepts and develop problem-solving strategies. Manipulatives such as Color Tiles have been proven to be crucial to children's understanding of mathematics. When children are actively involved with manipulatives, they intuitively develop an understanding of math concepts.

The activities in this book are designed for use with your mathematics curriculum in grades K-3. Color Tiles are used with every activity. The set of ***Color Tiles*** (LER 203) consists of durable 1-inch plastic squares in yellow, blue, red, and green. For teachers' use, the set of ***Overhead Color Squares*** (LER 478) are available in $^{3}/_{4}$-inch squares (set includes 5 each of 10 colors).

NCTM Standards

Curriculum and Evaluation Standards for School Mathematics, published by the National Council of Teachers of Mathematics (NCTM), was used as a guide in writing this book. Special attention was given to the following 12 standards:

- Math as Problem Solving
- Math as Communication
- Math as Reasoning
- Math Connections
- Estimation
- Number Sense and Numeration
- Concepts of Whole Number Operations
- Whole Number Computation
- Geometry and Spatial Sense
- Measurement
- Statistics and Probability
- Patterns and Relationships

Free Play

Introduce your children to Color Tiles by encouraging free play. Free play helps children become comfortable with the manipulatives. The free play time should precede the activities in this book. Begin free play by allowing children time to handle the tiles and play with them. However, give children some direction during free play, so they become accustomed to a structured setting. Here are some suggestions for children. Say: "Use your Color Tiles to:

- make a design;
- write your initials;
- build a house;
- create an animal;
- show a flower."

Observe how children react to the manipulatives, how they handle the manipulatives, and how they work in groups. This informal assessment can be very helpful later on. Children in your classroom more than likely will be at different levels of cognitive development. While assessing children, observe their levels of development. This will be a good indicator of children's readiness to move on.

USING THIS BOOK

• Color Tiles are packaged in a set of 400 pieces, which includes 100 red, 100 yellow, 100 blue, and 100 green tiles. Two sets of Color Tiles are recommended for your classroom.

• An overhead projector is helpful. You also should have overhead pens in 4 colors (yellow, blue, red, and green) to match the Color Tiles, but you can substitute shading or letters if necessary.

• Overhead Color Squares are recommended for many activities. If you do not have Color Squares, you can substitute shading or use letters when necessary. Note that the Color Squares are $^3/_4$-inch, not 1-inch squares.
Important note: Photocopy Tile Sheets at 75% when creating transparencies, <u>if</u> Overhead Color Squares are being used.

• Each activity consists of teaching notes and a corresponding reproducible activity page.

• When a transparency is recommended, it is noted in the Teaching Notes. These transparencies can be reproduced on acetate.

• Selected Answer Keys are found in the back of the book.

• Encourage children to be actively involved. When possible, ask children to demonstrate problems on the overhead, and call on them to answer questions.

• Cooperative learning is encouraged throughout the book. Working in groups and pairs can benefit children at different levels. Suggested group size is 4.

• Activities are sequenced developmentally under 3 levels: concrete, pictorial, and symbolic. Concrete activities give children hands-on experiences for a solid foundation in the mathematical areas explored. Pictorial activities represent the concepts with appropriate pictures. At the symbolic level, concepts are represented by symbols such as letters, numbers, and addition sentences.

• The reproducible Journal Page (page 49) gives children opportunities to draw and write about the math they have learned. In addition, it gives the activities closure and can be used as an evaluation tool. For young children, you may decide to focus on the drawing portion of the Journal Page. One approach could be to develop a set of guidelines that must be used every time in their journals. For example, encourage children to "Draw a picture of what you did today. Include all members of your group, the materials used, and a description of the activities in your drawing." Another approach could be to specify different instructions every time. For example, for the third activity on quilts, you may want to read a story about quilts, and then have children make up their own stories in their Journals. Another approach would be to have children design their own math problems on their Journal Pages.

• Chapter 8 presents games and puzzles that integrate the concepts covered in this book.

AREAS OF EMPHASIS:
Pattern Recognition and Sequencing

PATTERNS 1 Checkerboard Line Patterns

TEACHER MATERIALS

- Checkerboard Activity Master
- Overhead Color Squares
- Tile Sheet 1 Transparency

STUDENT MATERIALS

- Color Tiles (10 of each color per pair of children)
- Tile Sheet 1
- Colored pencils or crayons
- Journal Page

Concrete Activities

- Discuss patterns with children and the occurrence of patterns around them. Demonstrate a few simple patterns. Display the checkerboard and briefly explore patterns on the board.

- Cover all rows except the top row so that you have a checkerboard line.

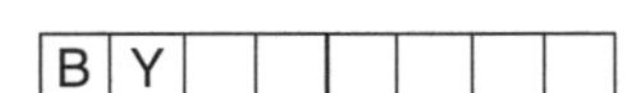

- Use blue and yellow overhead squares to build and demonstrate the pattern in the first row of the checkerboard.
- Pair up children. Ask children to copy and continue the pattern using blue and yellow tiles on one copy of Tile Sheet 1.
- Complete an overhead pattern for children to self-check. Tell children to leave their tiles on the first row.
- Ask children to build the pattern on the second row of the checkerboard. They should now have two rows of tiles on Tile Sheet 1.
- Discuss the two patterns. How are they similar? How are they different? How many of each row do you need to build the checkerboard? (Answer: 4)

Pictorial Activities

- Children remove each tile and then color in the squares with the matching color (blue or yellow).

Symbolic Activities

- On the bottom two rows of Tile Sheet 1, ask children to copy the patterns they discovered on the checkerboard. This time they need to use matching letters instead of colors. Depending on the level of your children, you may want to do this activity independently, in groups, or as a class.

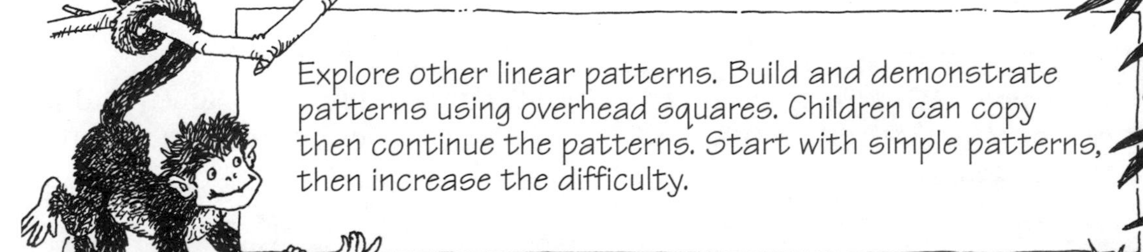

Name ____________________

TILE SHEET 1

AREAS OF EMPHASIS:
Pattern Recognition and Sequencing

PATTERNS

Checkerboard Patterns

TEACHER MATERIALS

- Checkerboard Activity Master
- Overhead Color Squares (2 Sets)
- Tile Sheet 2 Transparency
- Color Tile Grid Activity Master (optional)

STUDENT MATERIALS

- Color Tiles (10 of each color per pair of children)
- Tile Sheet 2
- Colored pencils or crayons
- Journal Page

Concrete Activities

- Display the checkerboard and discuss patterns on the board.

- Cover all rows except the top two so that you have a two-dimensional pattern.

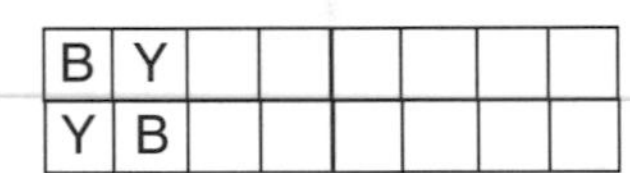

- Use Tile Sheet 2 Transparency on the overhead to build and demonstrate the pattern in the first two columns. Use 2 blue and 2 yellow squares.
- Pair up children. Children copy and continue the pattern using yellow and blue tiles and a copy of Tile Sheet 2 (Section A).
- Complete the overhead pattern. Children check to see that they have completed the pattern correctly.

Pictorial Activities

- Children remove and color in each square with the matching color (yellow or blue). Each student completes Section A.
- Children look at the third and fourth rows on the checkerboard. Children should copy the pattern using their colored pencils on Tile Sheet 2 (Section B). Note: Younger children may need to build the pattern with the tiles first.

Symbolic Activities

- Children look at the fifth and sixth rows of the checkerboard. On Section C of Tile Sheet 2, children put in the letters for the matching colors blue or yellow. Depending on the level of your children, you may want them to do the activity independently, in groups, or as a class while you walk them through it.

Explore other two-dimensional patterns. Build and demonstrate, patterns using overhead squares. Children copy, then continue the patterns. Increase the difficulty of patterns by varying the number and sequencing of tiles.

Name ____________________

TILE SHEET 2

A

B

C

AREAS OF EMPHASIS:
Pattern Recognition,
Sequencing, and Counting

PATTERNS 3 Quilts

TEACHER MATERIALS
- Quilts and quilt pictures
- Quilt Transparency
- Overhead Color Squares
- Color Square Grid Transparency
- Tile Sheet 3 Transparency

STUDENT MATERIALS
- Color Tiles (6 of each color per child)
- Tile Sheet 3 (2 per child)
- Colored pencils or crayons
- Journal Page

Concrete Activities

- Quilts are made up of blocks of patchwork designs, usually squares, which can be used alone or repeated to build an overall pattern. Display a quilt or pictures you have brought in and find the square blocks.

R	Y	R
Y	Y	Y
R	Y	R

- The Quilt Transparency displays a quilt made of 16 blocks. Each block is made of 9 squares. On the overhead, color in the first block to further emphasize the pattern. Ask: "How many blocks are there?" (Answer: 16) "Do they all follow the same pattern?" (Answer: Yes)

B	R	Y	B
R	Y		

- Use Tile Sheet 3 Transparency. Ask: "How many Color Tiles make up the block?" (Answer: 16) Use the overhead squares or overhead pens to begin the 3-color pattern. Children can work in pairs or individually. Ask children to copy and continue the pattern using Color Tiles and the top half of their Tile Sheet 3. Complete the pattern on the overhead for children to self-check. Discuss the overall pattern. Ask: "Were you surprised to see the design?"

- Ask students to create their own block designs using the bottom half of Tile Sheet 3.

Pictorial Activities

- Ask for volunteers to demonstrate their patterns on the overhead. You will need overhead pens and the Color Square Grid Transparency.

Symbolic Activities

- On the overhead build a 16-tile block. Ask children to document this pattern by writing in the corresponding letters.

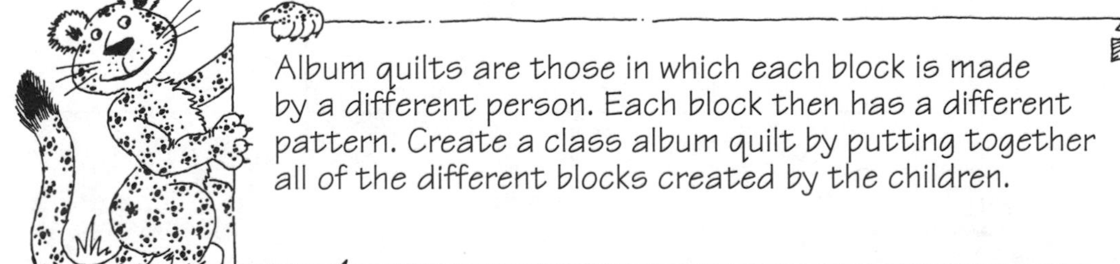

Name ___________________

TILE SHEET 3

Copy and continue the pattern.

Create your own pattern.

AREAS OF EMPHASIS:
Counting, Estimating, and Ordering

COUNTING

Bagging Tiles

TEACHER MATERIALS

- 3 resealable plastic bags: one with 6 tiles labeled A, one with 19 tiles labeled B, one with 20 tiles labeled C.
- Paper bag
- Overhead Color Squares (2 Sets)
- Tile Sheet 4 Transparency

STUDENT MATERIALS

- Color Tiles (25 per group)
- Resealable plastic bags (6 per group)
- Tile Sheet 4
- Paper bags (optional)
- Colored pencils or crayons
- Journal Page

Concrete Activities

- Discuss estimation with children. Ask: "What is estimation? When do you use estimation?" Then ask children for examples of estimation: There are about 5 cookies left, I have about 10 marbles at home, etc.

- Before class place your 3 plastic bags labeled A, B, and C inside a paper bag. From your paper bag pull out the plastic bag with the 6 tiles (A) and one of the other plastic bags (B or C). Hold up the bags for everyone to see. Ask: "Which bag has more tiles, B or C? How can you tell? Give me an estimate for howmany tiles are in each bag."

- Return the bag with 6 tiles to the paper bag and pull out the other bag. "Can you tell, by just looking, which bag has more, B or C?" (Answer: No) Sometimes estimates are good for telling which has more and which has less, but sometimes we need more information.

- Have children work in groups. Ask group members to place 1 tile in the first plastic bag, 2 tiles in the second plastic bag, 3 tiles in the third plastic bag, 4 tiles in the fourth plastic bag, and 5 tiles in the fifth plastic bag. Children then place the bags in order from 1 to 5.

Pictorial Activities

- On Tile Sheet 4, ask children to draw the number of tiles in their ordered bags.

Symbolic Activities

- On Tile Sheet 4, children write the matching number beneath each bag.

Groups place their five plastic bags inside a paper bag. In another plastic bag, children place a random number of tiles between 1 and 10 and add the plastic bag to the paper bag. They trade bags with another group. Each group then orders their new set of bags to find which plastic bag was added last.

Name ______________________

TILE SHEET 4

Draw the tiles that belong in each bag (from 1 to 5).
Write the number of tiles in each bag on the lines below.

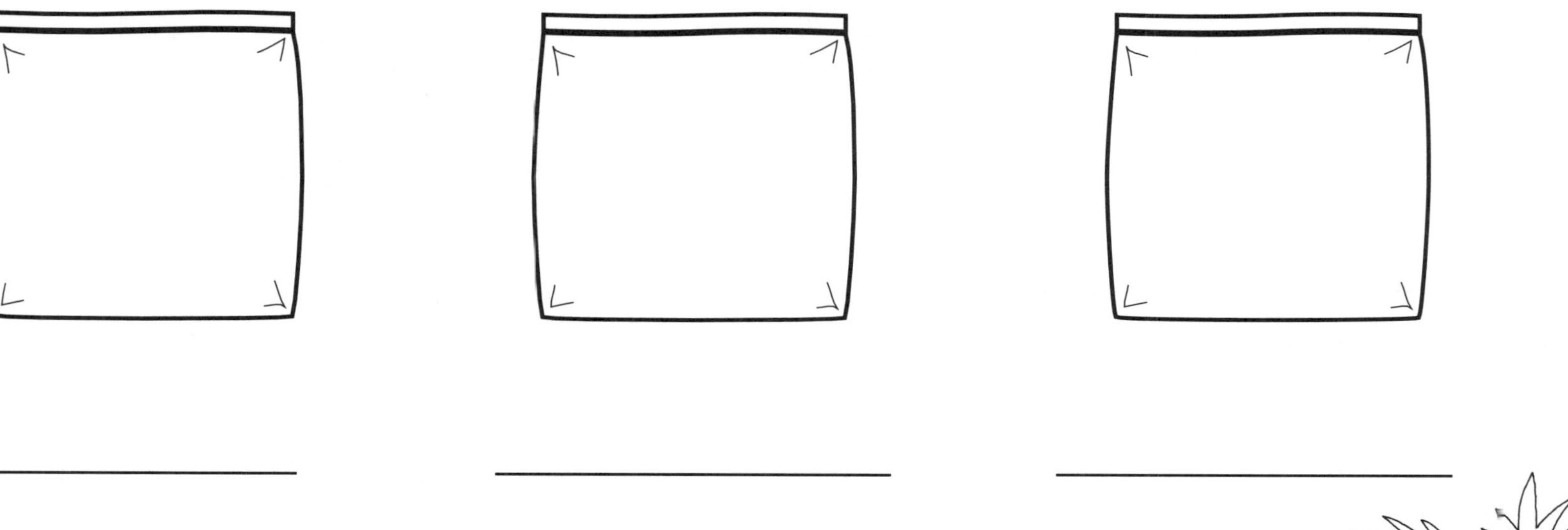

COUNTING 2

Calendar Count

TEACHER MATERIALS
- Wall Calendar
- Overhead Color Squares
- Tile Sheet 5 Transparency

STUDENT MATERIALS
- Color Tiles (31 yellow, 1 blue, 5 red, 5 green per group)
- Tile Sheet 5
- Colored pencils or crayons
- Journal Page

Concrete Activities

- Ask children to name the current month, day, and year.

- Display a calendar for children to see and verify the name of the month, day, and date. Point out how the days are numbered in order.

- Have children work in groups for this activity. Ask students to fill in the name of the month and year on their calendars. Use Tile Sheet 5 Transparency. Point out where 1 would be on the calendar. Ask children to point to the same place on their copies of Tile Sheet 5. Explain: "We are going to put yellow tiles on the calendar. Count along with me." Place a yellow tile where the first day of the month would fall and count "1." All children should count along. Place a yellow tile and count out each day of the month. Ask: "How many yellow tiles did we use? How many days are in this month?"

- Ask children to replace today's date with a blue tile (as you do it on the overhead). "How many yellow tiles are left?" Have children replace all Saturdays with red tiles. "How many red tiles are there? How many Saturdays are there?" Have children replace all Sundays with green tiles. "How many green tiles are on the calendar? How many Sundays are there?"

Pictorial Activities

- Ask children to remove each tile and color in the calendar with the corresponding color (yellow, blue, red or green). Children should color their individual calendars to match the group calendar.

Symbolic Activities

- Have children look at their calendars and fill in the dates starting with 1. Have the month's calendar available for reference.

What would happen if you place 1 tile on the first day, 2 tiles on the second day, 3 tiles on the third day, and so on? How many tiles do you think you would need? 30? 60? 100? Take guesses and record them. (Answer: For a 30-day month, you would need 465 tiles.)

TILE SHEET 5

Name ____________________

Month ____________________ **Year** ____________

AREAS OF EMPHASIS:
Place Value, Counting, Renaming, and One-to-One Correspondence

Jaguar-Count

TEACHER MATERIALS
- Overhead Color Squares
- Jaguar-Ones Transparency
- Jaguar-Tens Transparency
- Tile Sheet 6 Transparency

STUDENT MATERIALS
- Color Tiles (20 per pair of children)
- Jaguar-Ones Activity Master (10 per student)
- Jaguar-Tens Activity Master (2 per student)
- Tile Sheet 6
- Colored pencils or crayons
- Journal Page

Concrete Activities

- Children work in pairs. Ask children to show you 2 yellow tiles, then 5 yellow tiles, then 8 yellow tiles.
- Distribute Jaguar-Ones. Ask children to show you 2 Jaguar-Ones and place them on top of the 2 yellow tiles. Have children do the same with 5 Jaguar-Ones, and then 8 Jaguar-Ones.
- Distribute Jaguar-Tens. Ask children to show you 10 jaguars using Jaguar-Ones, then show you 10 jaguars using a Jaguar-Ten strip. Compare the two sets. How are they different? (Answer: Jaguar-Tens are attached; Jaguar-Ones are not attached.) How are they same? (Answer: Both sets contain 10 jaguars.) Demonstrate comparison on the overhead.
- Ask children to show you 13 jaguars using Jaguar-Ones only. Then ask them to show you 13 jaguars using both Jaguar-Tens and Jaguar-Ones. Compare the two sets. Ask: "What is the same? What is different?"
- Place 1 Jaguar-Ten strip and 4 Jaguar-Ones on the overhead. Ask children to name the value. (14) Repeat several times.

Pictorial Activities

- On Tile Sheet 6, Problem 1, each student colors in the number of tiles that corresponds to each number.

Symbolic Activities

- On Tile Sheet 6, Problem 2, each student writes the number that corresponds with the number of tiles shown. Use Tile Sheet 6 Transparency to demonstrate the problems. The last problem involves 2 Jaguar-Ten strips as a challenge.

Ask children to show you larger numbers such as 21, 33, and 48, using Jaguar-Tens and Jaguar-Ones. Show large numbers using Jaguar-Tens and Jaguar-Ones, and have children give the matching numbers.

Name ______________________

TILE SHEET 6

1. Color the squares yellow to show each number.

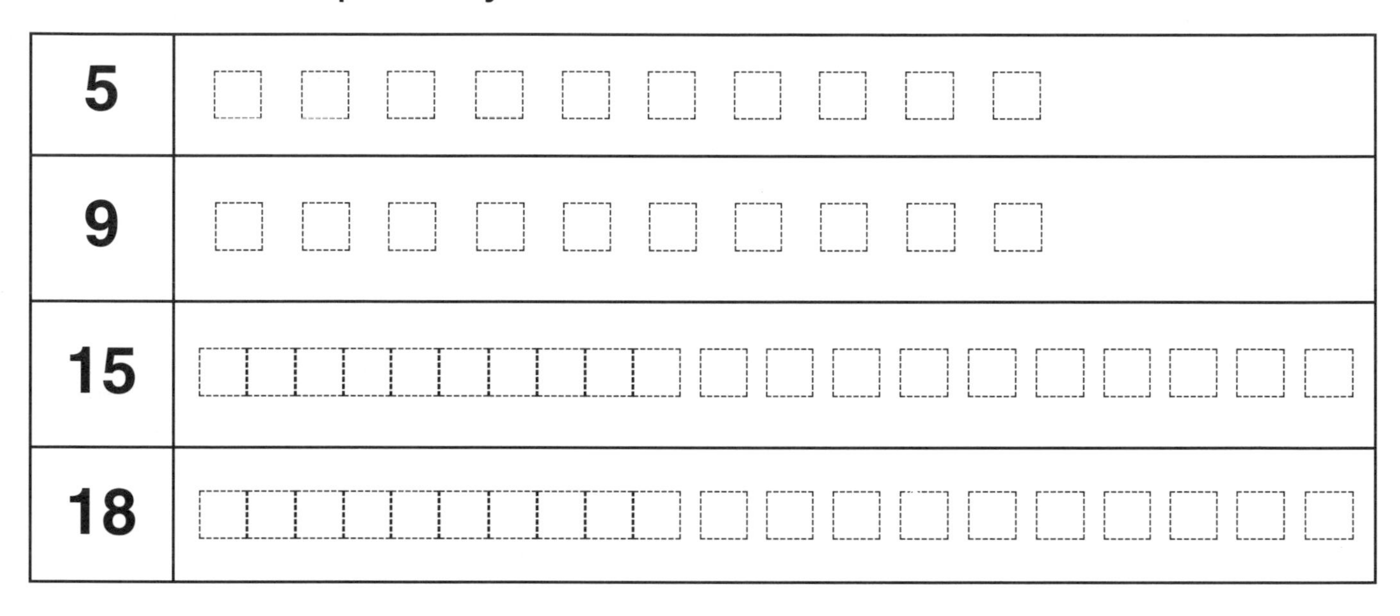

2. Give the numbers that stand for the tiles shown.

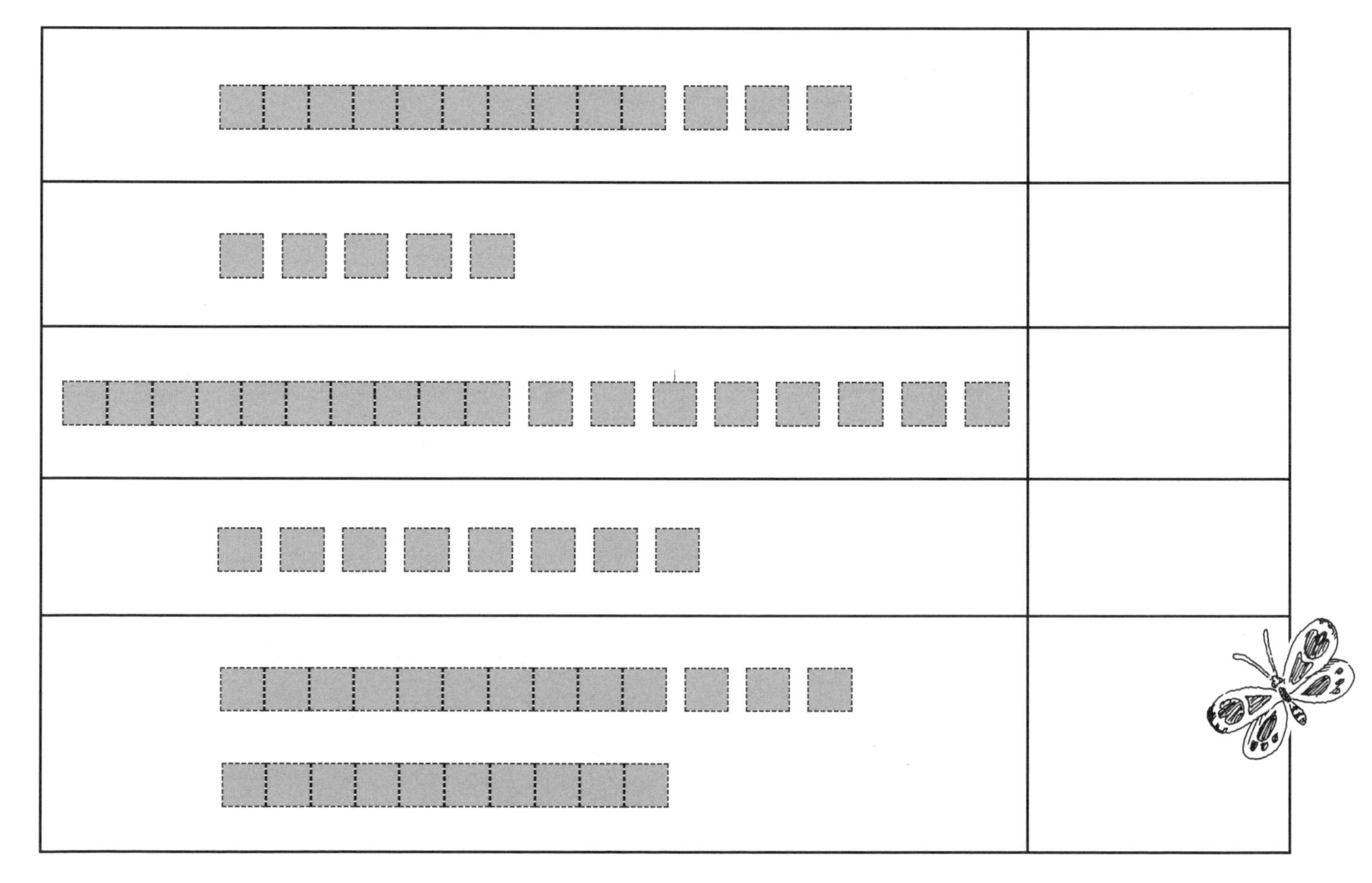

PLACE VALUE 2

AREAS OF EMPHASIS
Place Value, Counting, and Sorting

Bags of Tiles

TEACHER MATERIALS
- Overhead Color Squares (2 sets)
- Tile Sheet 7 Transparency

STUDENT MATERIALS
- Color Tiles (at least 15 per child)
- Resealable plastic bags (4 per group)
- Tile Sheet 7
- Colored pencils or crayons
- Journal Page

Concrete Activities

- Distribute tiles and plastic bags. Ask each student to count out 10 yellow tiles and place them in a plastic bag. Ask: "How many tiles are there if there is one bag? (10) 2 bags? (20) 3 bags? (30) Four bags? (40)"

- Use Tile Sheet 7, Problem 1. Children work in groups. There should be at least 4 children in each group so you have a total of 40 bagged tiles (4 bags) and 20 single tiles. Demonstrate the first problem. Write the number 12 on the board or overhead. How can you use your bags and tiles to show this number? (Answer: Show 1 bag and 2 tiles.)

- Write the numbers 8, 11, 14, and 18 on the board or overhead. As a group, children decide how to show each of the numbers using their bags and tiles. Ask for a volunteer from each group to demonstrate one of the numbers using their bags and tiles on the overhead.

Pictorial Activities

- Use Tile Sheet 7, Problem 1. Children draw the bags and tiles for each number. Encourage children to draw the most efficient combination of bags and tiles when possible. For example, 15 single tiles for the first problem would not be the most efficient answer.

Symbolic Activities

- Children complete Tile Sheet 7, Problem 2. Use Tile Sheet 7 Transparency to review the problems.

In your head can you tell me how many tiles are in 30 bags? (300) 31 bags? (310) 32 bags? (320) Do you see any patterns? What are they? How many tiles do you have if you have 39 bags and 4 tiles? (394). 173 bags and 9 tiles? (1739). Build a table to help you see patterns.

Name ____________________

TILE SHEET 7

How many tiles are in each bag? ______

1. Draw the bags and tiles that stand for the numbers in the following table. The first one is finished.

Number	Bags	Tiles
15	▢	▦ ▦ ▦ ▦ ▦
19		
22		
32		
34		
47		

2. Answer the following questions.

Sarah has 1 bag and 8 tiles.
How many tiles does Sarah have? ____________

Anthony has 2 bags and 7 tiles.
How many tiles does Anthony have? ____________

Taylor has 3 bags and 5 tiles.
How many tiles does Taylor have? ____________

Kyle has 4 bags and 6 tiles.
How many tiles does Kyle have? ____________

AREAS OF EMPHASIS:
Plave Value, Renaming, Counting, and Sorting

PLACE VALUE

Trading Tiles

TEACHER MATERIALS
- Overhead Color Squares (2 Sets)
- Tile Sheet 8 Transparency
- Transparency 1
- 1 blue and 10 yellow Color Tiles
- Resealable plastic bag

STUDENT MATERIALS
- Color Tiles (10 of each color per pair of children)
- Tile Sheet 8
- Journal Page
- Colored pencils or crayons

Concrete Activities

- Explain that one yellow tile is the same as one (1). Place 10 yellow tiles in a plastic bag. Show the bag and 1 more yellow tile and ask for the value. (Answer: 11)

- Show a blue tile. Then explain: “I want to exchange tiles. The rule for today is that I can exchange one blue tile for this bag of yellow tiles.” (Let children know that the rules can change from day to day. This will avoid confusion in subsequent problems.) “Imagine this yellow tile is a penny and the blue tile is a dime. What is the value of this blue tile?” (10¢) Show 3 blue tiles. “What is the value of 3 blue tiles?” (30¢)

- Complete Transparency 1 as a whole class activity. Place a blue and yellow tile on the first row, then ask for the value. (11) On the next row, place 1 blue and 2 yellow tiles on the squares. Ask for the value of the combination. (12) Continue giving combinations and asking for the values.

- Write 15 on the chalkboard or overhead. Direct children by saying: “Use the least number of color tiles to show 15.” Here you may need to discuss value (15) versus number (6) of tiles. Children use tiles to represent the number (one blue and five yellow tiles). Write several more examples for children.

Pictorial Activities

- Children complete Tile Sheet 8, Problem 1. Children color the tiles to show the combinations needed for each number.

Symbolic Activities

- Children complete Tile Sheet 8, Problem 2. Children match each combination with its number. Use Tile Sheet 8 Transparency to review the problems.

Investigate hundreds and thousands using Color Tiles. Ask: “Exchange 1 green for 10 blues. What is the value of 3 green, 2 blue, and 5 yellow tiles? (325) Exchange 1 red for 10 greens. What is the value of 7 red, 8 green, 2 blue, and 6 yellow tiles? (7,826) Show me 4,008 using color tiles.”

Name ____________________

TILE SHEET 8

Yellow = 1 Blue = 10

1. Color the squares blue or yellow to show each number.

18	☐	☐	☐	☐	☐	☐	☐	☐	☐			
24	☐	☐	☐	☐	☐	☐						
39	☐	☐	☐	☐	☐	☐	☐	☐	☐	☐	☐	☐
41	☐	☐	☐	☐	☐							
52	☐	☐	☐	☐	☐	☐	☐					

2. Draw a line from each tile combination to its matching number.

B B Y	26
Y Y B	33
B Y B Y B Y	21
Y B Y Y B Y B	34
Y Y Y Y Y Y B B	12

ADDITION AND SUBTRACTION 1

AREAS OF EMPHASIS: Addition, Counting, Sorting, and Inclusion

The Whole and Its Parts

TEACHER MATERIALS

- Overhead Color Squares (2 Sets)
- Tile Sheet 9 Transparency

STUDENT MATERIALS

- Color Tiles (10 yellow, 6 blue, 4 green, and 2 red per child)
- Tile Sheet 9
- Colored pencils or crayons
- Journal Page

Concrete Activities

- Say to children: "Group your tiles according to color. How many yellow tiles are there? (10) How many blue? (6) Green? (4) Red? (2) Are there more yellow tiles or more blue tiles? (Yellow) Are there more red tiles or tiles, in general? (Tiles) Are there more yellow tiles or tiles altogether?" (Tiles)

Pictorial Activities

- Tile Sheet 9, Problem 1. Ask children to fill in the middle column by drawing and coloring the number of tiles for each color. (10 yellow, 6 blue, 4 green, and 2 red)

Symbolic Activities

- Tile Sheet 9, Problem 1. Ask children to fill in the final column by writing the number of tiles shown.
- Ask children to complete Problem 2 by adding the tiles indicated. Each problem can be modeled with overhead tiles. For example, for Y and B, place the 6 blue tiles next to the 10 yellow tiles, then count to find the sum. (16) Children fill in the total number of tiles on their tile sheets.

Y and B give you 16. What other combination also gives you 16? (B and Y) List other combinations that when put together give the same number. (Y + G and G + Y, Y + R and R + Y, B+ G and G + B, B + R and R + B, G + R and R + G). Ask: "When are the sums equal?" (When the same colors are used.)

Name ____________________

TILE SHEET 9

1. Color of Tile	Draw the Tiles	How many?
Yellow		
Blue		
Green		
Red		

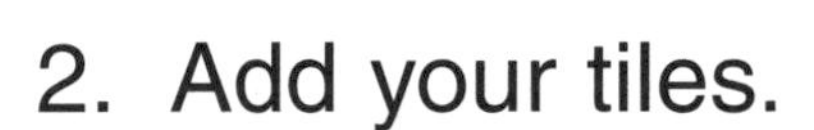

2. Add your tiles.

Y and B ______

Y and G ______

Y and R ______

B and Y ______

B and G ______

B and R ______

G and Y ______

G and B ______

G and R ______

R and Y ______

R and B ______

R and G ______

ADDITION AND SUBTRACTION 2

AREAS OF EMPHASIS:
Addition and Counting

Addition Sentences

TEACHER MATERIALS
- *Overhead Color Squares*
- *Tile Sheet 10 Transparency*
- *Transparency 2*

STUDENT MATERIALS
- *Color Tiles (red and blue, 5 of each per child)*
- *Tile Sheet 10*
- *Colored pencils or crayons*
- *Journal Page*

Concrete Activities

- Use Transparency 2. Place one red and one blue square on the overhead. Say to students: "We are going to be using only the red and blue tiles today. How many reds do I have?" (1) Write 1 in the table. "How many blues do I have?" (1) Write 1 in the table."How many do I have altogether?" (2) Write 2 in the table. "Let's write an addition sentence for this problem: 1 + 1 = 2 ." Write the addition sentence on the line provided below the squares. "Can you think of a different combination?" 1 blue and 1 red is the same as 1 red and 1 blue. "How about two of the same color?" Place two reds on the squares, fill in the table, and write the addition sentence. (2 + 0 = 2) Do the same for two blues.

- Use the bottom half of Transparency 2. Ask for suggestions about how to place red and blue tiles on the three squares put together. Emphasize that you are looking for different combinations; 1 red (R) and 2 blue (B) is the same as 2B and 1R. Guide children to discover that there are 4 possible combinations. (1R and 2B, 2R and 1B, 3B, and 3R) Complete the table and write the addition sentences. Ask, "What is similar in the table and in the addition sentences?" (If you add + and = signs to the table, you have your addition sentences).

- Use Tile Sheet 10. Ask children to use red and blue tiles to find the possible combinations for 4 squares and 5 squares. Ask children to work in groups to solve the problems.

Pictorial Activities

- Use Tile Sheet 10. Ask children to color in the squares to show the different combinations possible for 4 and 5 squares.

Symbolic Activities

- Use Tile Sheet 10. Students should complete the tables by filling in the corresponding numbers. Point out the addition sentences within each table. Use Tile Sheet 10 Transparency to check answers as a whole class. Ask group members to come up to the overhead to fill in the answers.

Ask: "Do you see any patterns in the table? Can you predict how many addition sentences I will have if I have 6 squares? (7) 10 squares? (11) 15 squares? (16)"

Name ______________________

TILE SHEET 10

Use red and blue tiles. Find the different combinations.

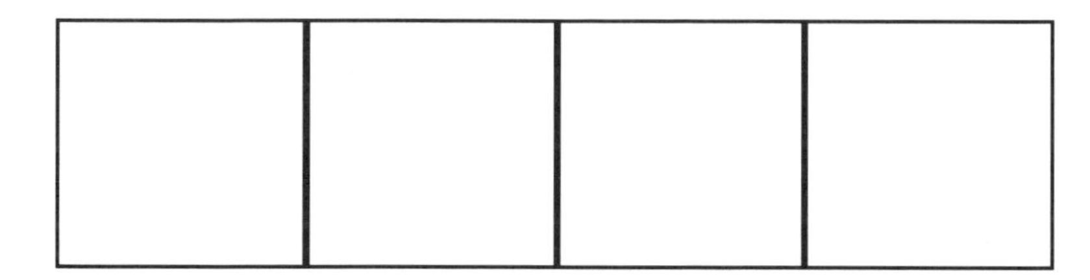

Draw your combinations here.

How many R?	How many B?	How many altogether?
+	=	
+	=	
+	=	
+	=	
+	=	

Draw your combinations here.

How many R?	How many B?	How many altogether?
+	=	
+	=	
+	=	
+	=	
+	=	
+	=	

ADDITION AND SUBTRACTION 3

AREAS OF EMPHASIS:
Addition, Subtraction, and Counting

Give Me Your Tiles

TEACHER MATERIALS
- Overhead Color Squares
- Tile Sheet 11 Transparency

STUDENT MATERIALS
- Color Tiles
- Tile Sheet 11
- Colored pencils or crayons
- Journal Page

Concrete Activities

- Draw large + and = signs on the board. Ask for three volunteers to come to the front of the room and stand with the signs between them.

  Give the first volunteer (LaTanya) 2 tiles, and the second (Sharon) 4 tiles. Explain: "LaTanya and Sharon will now give their tiles to Jaime. Ask Jaime to count his tiles. Jaime now has 6 tiles." Repeat this activity with different volunteers and different addends for additional reinforcement. Note: Some children may have difficulty seeing that 2 + 4 is the same as 4 + 2. This is an excellent opportunity to demonstrate this.

- Ask for three volunteers. Give the last volunteer (Claudia) 8 tiles. Say: "Claudia has 8 tiles and Roberto gave her 3 of those tiles. Let's find out how many of those tiles Monique gave her." The child with 8 tiles gives the first student 3 tiles and the rest of the tiles to the second child. The second child counts the tiles to see how many she has. (5) Repeat this activity with different volunteers and different variations of tiles.

Pictorial Activities

- Have children complete Tile Sheet 11 by drawing in the missing tiles. Allow children to use Color Tiles to solve the problems. Note: Problem 6 is open-ended and has many solutions. Use Tile Sheet 11 Transparency to review the problems.

Symbolic Activities

- Ask children to go back to each of the problems on Tile Sheet 11 and write the addition sentence for each.

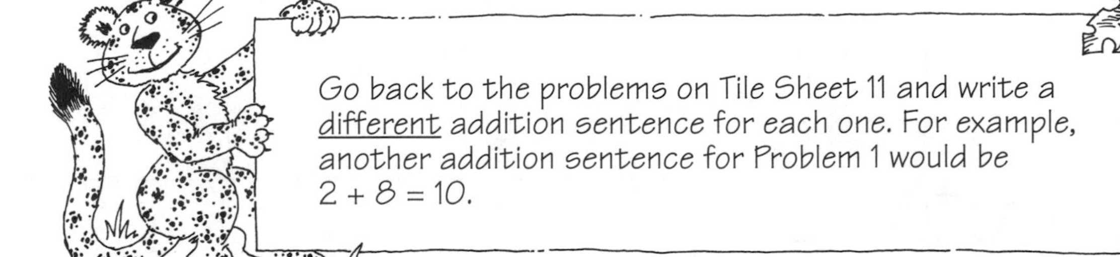

Name ____________________

TILE SHEET 11

Draw in the missing tiles.

1.

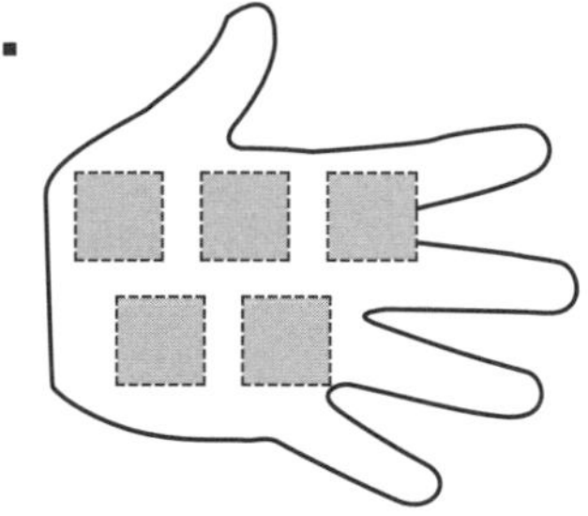 + 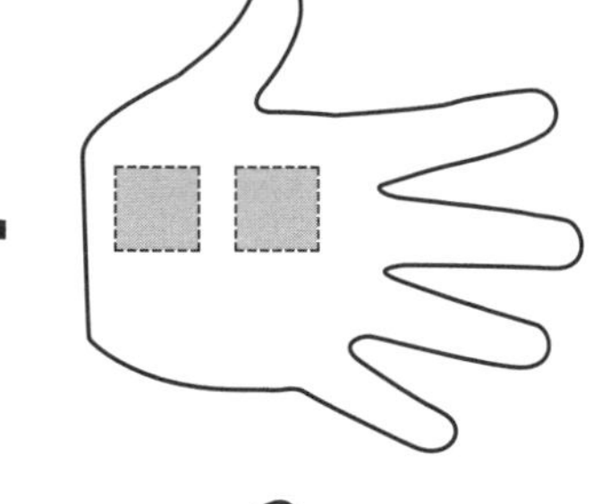 = 

2.

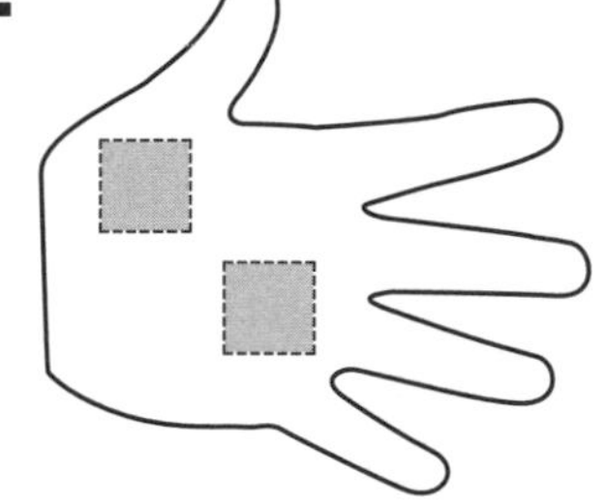 + 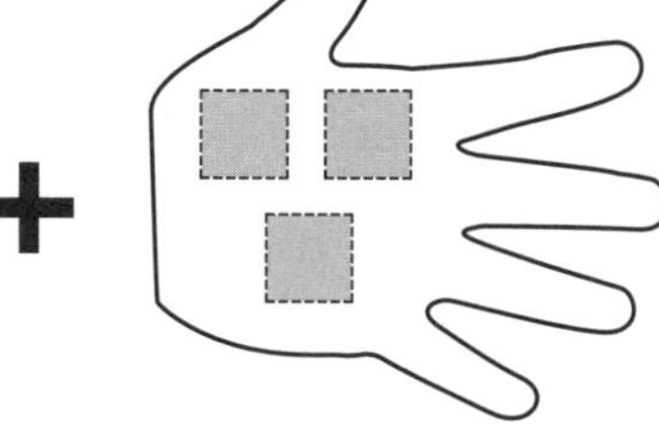 = 

3.

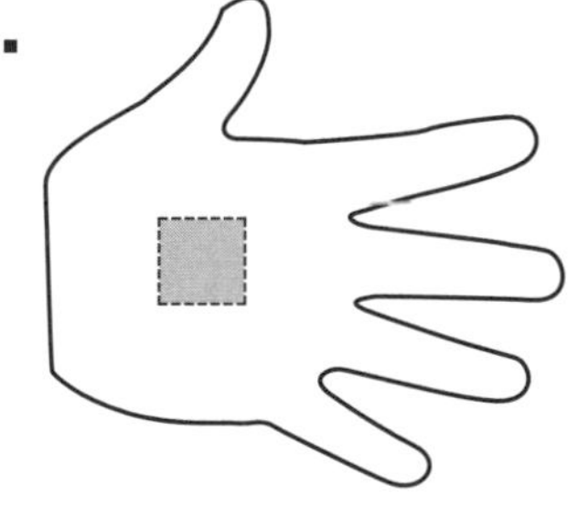 +  = 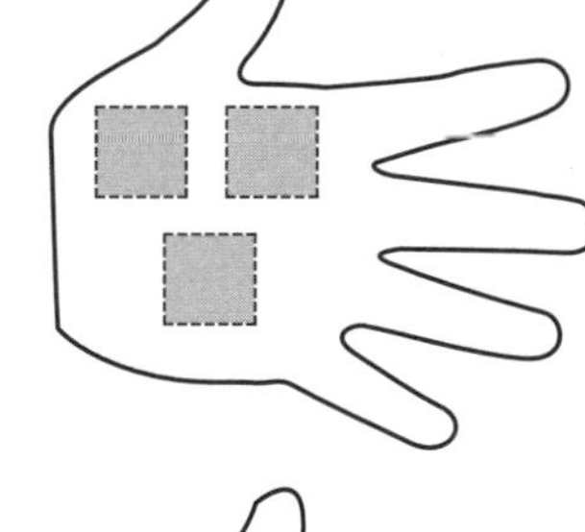

4.

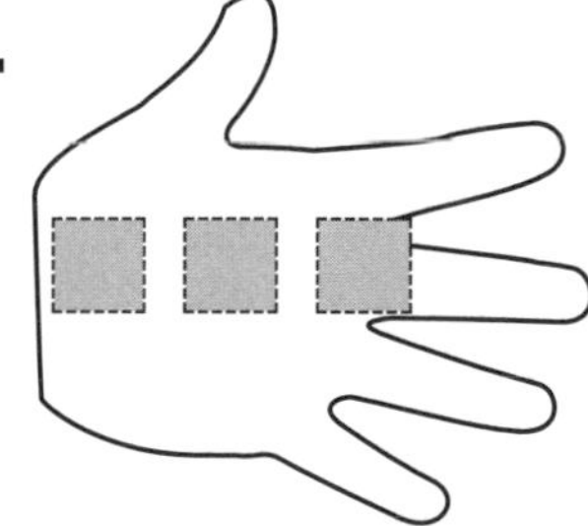 + 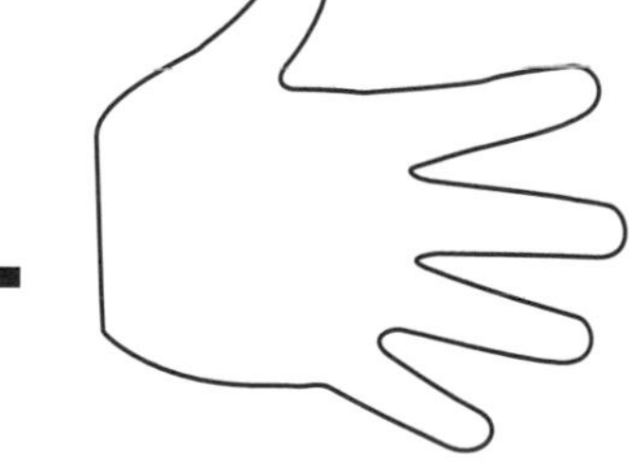 = 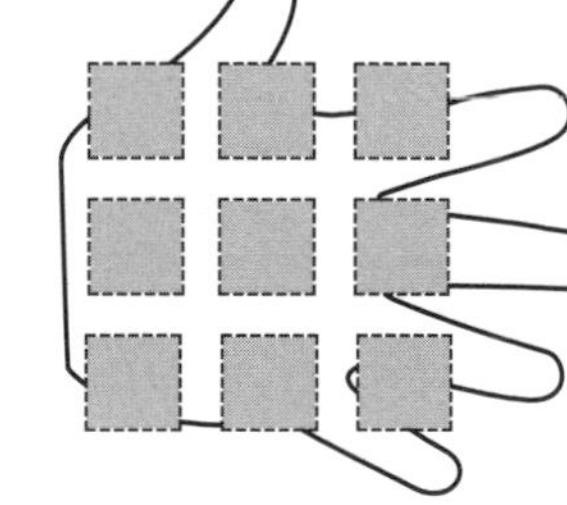

5.

 + 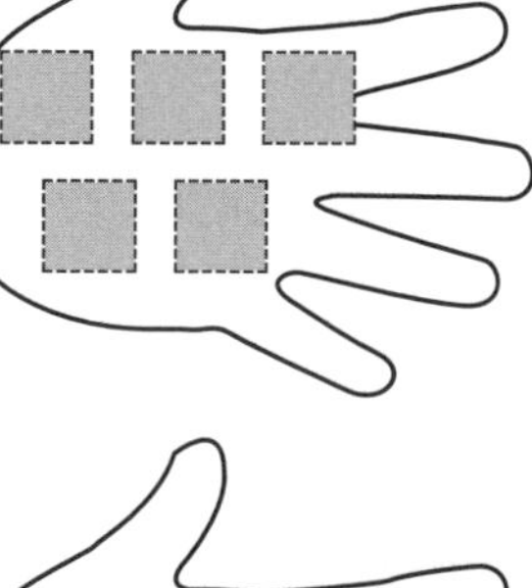 = 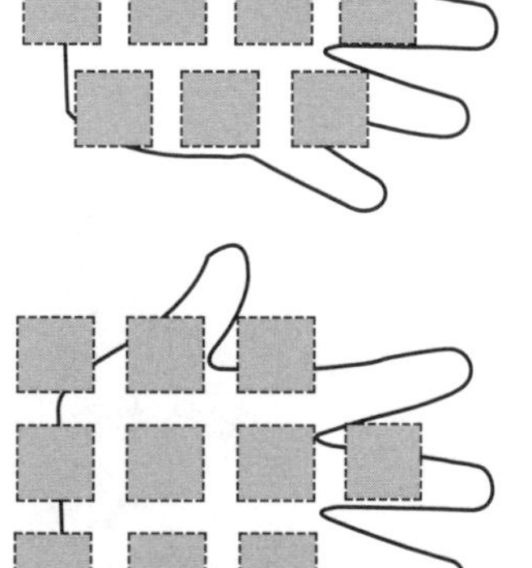

6.

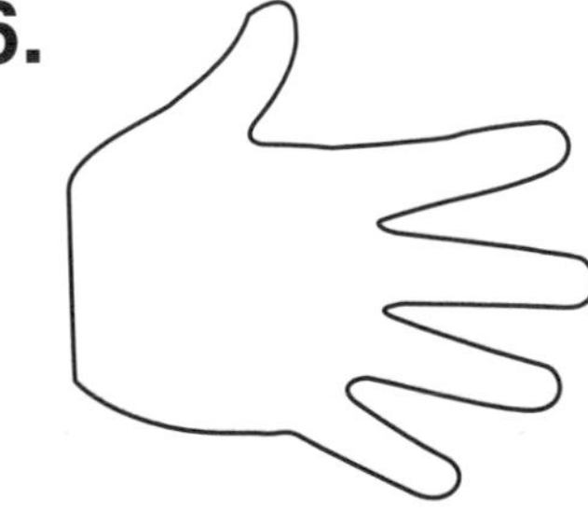 

ADDITION AND SUBTRACTION 4 Subtraction Action

AREAS OF EMPHASIS:
Subtraction, Addition, and Counting

TEACHER MATERIALS
- *Color Tiles*
- *Tile Sheet 12 Transparency*

STUDENT MATERIALS
- *Color Tiles*
- *Tile Sheet 12*
- *Colored pencils or crayons*
- *Journal Page*

Concrete Activities

- Ask 5 children to stand together at the front of the room. Place children into two groups, a group of 3 and a group of 2. Say: "If I have 5 children in a group and I take away, or subtract, 2 students, how many will I have left? (3) I have 3 children in this group; how many are in the group I subtracted? (2) If I put the two groups back together or add them, how many do I have?" (5) Repeat Activity 1, this time grouping the children into groups of 1 and 4.

- Place 5 tiles on the overhead. Ask: "How many tiles do I have?" (5) Separate them into a group of 3 and a group of 2. "If I have 5 tiles and I subtract 2 tiles (cover the group with two tiles) how many do I have left?" (3) This can be written as 5 - 2 = 3. Relate the written numbers to the tiles and discuss the minus sign as the action of taking away. "How many tiles are there if I put the groups together, or add?" (5) Join and count to check the answer. Have a volunteer write down the addition sentence on the overhead (3 + 2 = 5 or 2 + 3 = 5). Compare the addition sentence to the subtraction sentence, and discuss how the equations are related.

- Put the five tiles together. Ask: "Can you think of another way to arrange five tiles into two sets?" Use children's suggestions to repeat Activity 2 to develop subtraction facts from 5-0 to 5-5.

Pictorial Activities

- Ask children to complete the table on Tile Sheet 12 (Problem 1), using Color Tiles if necessary.

Symbolic Activities

- Have children complete Problem 2 on Tile Sheet 12. Use Tile Sheet 12 Transparency to review the problems. Compare and discuss the answers to Problem 2 and the table. "Do you see any patterns? What are they?"

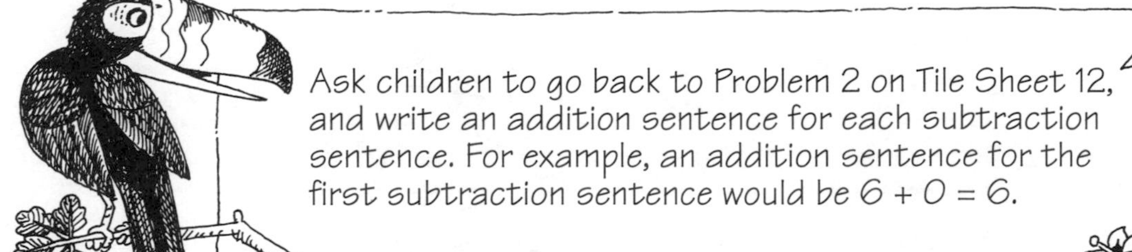

Ask children to go back to Problem 2 on Tile Sheet 12, and write an addition sentence for each subtraction sentence. For example, an addition sentence for the first subtraction sentence would be 6 + 0 = 6.

Name ____________________

TILE SHEET 12

1. Draw the tiles to show the different ways you can group 6. The first two are done for you.

Group tiles like this	First group	Second group
6 and 0	□□□□□□	
5 and 1	□□□□□	□
4 and 2		
3 and 3		
2 and 4		
1 and 5		
0 and 6		

2. Complete the following problems.

6 - 0 = ____

6 - 1 = ____

6 - 2 = ____

6 - 3 = ____

6 - 4 = ____

6 - 5 = ____

6 - 6 = ____

ADDITION AND SUBTRACTION

AREAS OF EMPHASIS:
Addition and Counting

Tile Sum

TEACHER MATERIALS
- *Overhead Color Squares (2 Sets)*
- Tile Sheet 13 Transparency

STUDENT MATERIALS
- Blue and Yellow Color Tiles
- Tile Sheet 13
- Colored pencils or crayons
- Journal Page

Concrete Activities

- Show a group of 4 yellow tiles and a group of 3 yellow tiles. Ask: "If I have 4 yellow tiles and add 3 yellow tiles, how many do I have?" (7) Put the two groups together and count to verify the answer. "I can write an addition sentence that looks like this: 4 + 3 = 7." Do several more problems as needed to reinforce.

- "The rule for today is that I can exchange 10 yellow tiles for 1 blue tile." Emphasize that this rule is only valid for that day. Show 1 blue tile and 4 yellow tiles. "How many do I have here?" (14) Do several more problems to review place value as needed.

- Show 1 blue tile and 2 yellow tiles.
  "What is the value of the combination?" (12) | B | Y | Y |
  Below that show 1 blue tile and 3 yellow tiles.
  "What is the value of the second combination? (13) | B | Y | Y | Y |
  I want to find out the value of all the tiles. To do this I can just add the tiles that have the same color." Group the blue tiles together, then the yellow tiles.

  What is the value of the tiles? (25) | B | B | | Y | Y | Y | Y | Y |

Pictorial Activities

- Tile Sheet 13, Problem 1 asks children to add blue and yellow tiles and write the results. Encourage children to use Color Tiles if necessary.

Symbolic Activities

- Problem 1 also asks children to look at the sum of numbers that correspond to the tile problems. Have children complete Problem 2. Encourage them to use Color Tiles, but only if necessary.

Give children addition problems that involve regrouping. First, they should use tiles to solve the problem. They should then use the fewest number of tiles to show their answers. Replace groups of ten yellow tiles with one blue tile.

Name ____________________

TILE SHEET 13

Yellow = 1	Blue = 10

1. Add the blue and yellow tiles below. Draw the tiles to show the answer. Then write the answer to the problem.

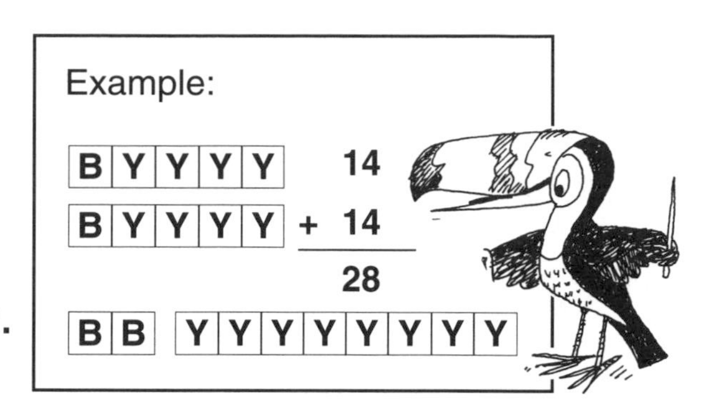

a)

B	B	Y

\+

B	Y	Y

____

$$\begin{array}{r} 21 \\ +\ 12 \\ \hline \end{array}$$

b)

B	B	B	B	Y	Y

\+

B	B	B	Y	Y	Y

____

$$\begin{array}{r} 42 \\ +\ 33 \\ \hline \end{array}$$

2. Solve each of the following problems.

$$\begin{array}{r} 22 \\ +\ 63 \\ \hline \end{array} \qquad \begin{array}{r} 38 \\ +\ 51 \\ \hline \end{array} \qquad \begin{array}{r} 75 \\ +\ 13 \\ \hline \end{array}$$

AREAS OF EMPHASIS:
Measurement, Estimating, Counting, and Comparison

MEASUREMENT 1 Tile Measure

TEACHER MATERIALS
- Color Tiles
- Tile Sheet 14 Transparency

STUDENT MATERIALS
- Color Tiles (divided equally among groups)
- Tile Sheet 14
- Colored pencils or crayons
- Journal Page

Concrete Activities

- Draw a line 7 inches long on the overhead. Ask: "How many tiles long do you think this line is?" Write students' estimate next to the line. Then measure the line using Color Tiles. Place the Color Tiles edge-to-edge next to the line to measure. Emphasize the importance of lining up the first tile with the beginning of the line. Compare the actual length with the estimated lengths, and circle the estimate that is the closest. Discuss how estimates are useful in getting an idea of the lengths of objects.

- Draw a 6-inch line and a 5-inch line on the overhead. Ask: "Which line do you think is bigger?" Have children estimate the length of each line. As a class, decide on one estimate for each line.

- Ask volunteers to measure each line with tiles on the overhead.

Pictorial Activities

- Ask children to draw a sketch of the lines and the correct number of tiles as shown on the overhead.

Symbolic Activities

- Children work in groups to complete Tile Sheet 14. In this activity, children collect objects, estimate lengths, find the lengths using Color Tiles, and then express the actual lengths with numbers.

- As you discuss the activity on Tile Sheet 14, remind children that each tile has a length of 1 inch. Ask: "Did anyone realize this? Which object did not have to be measured?" (Answer: ruler)

Make a list of things children can find at home and ask volunteers to bring the objects to school Estimate their lengths. Look at the first set of estimates. Do students want to change them? Measure the objects with Color Tiles. Compare the lengths to the estimates.

Name ____________________

TILE SHEET 14

1. Collect the objects listed in the table below. The last one is a free choice.
2. Estimate the length of each object.
3. Now use your Color Tiles to find the length of each object. Measure to the nearest tile.
4. Compare your estimates with your actual lengths.

Object	Estimate	Actual Length
Pencil	tiles	tiles
Paper Clip	tiles	tiles
Book	tiles	tiles
Desk	tiles	tiles
Calculator	tiles	tiles
Ruler	tiles	tiles
	tiles	tiles

AREA OF EMPHASIS:
Measurement, Estimating, Counting, and Place Value

MEASUREMENT 2

Longer Measure

TEACHER MATERIALS
- Color Tiles
- Tile Sheet 15 Transparency

STUDENT MATERIALS
- Color Tiles
- Tile Sheet 15
- Jaguar-Ones (Activity Master 5)
- Jaguar-Tens (Activity Master 6)
- Colored pencils or crayons
- Journal Page

Concrete Activities

- Copy Jaguar-Ones and Jaguar-Tens on blue, green, red, and yellow paper. Give each group 10 Jaguar-Ones and 3 Jaguar-Tens strips of each color.

- Ask children to estimate the length of the classroom. As a class agree on one estimate. Write the estimate on Tile Sheet 15 Transparency. Children can copy the estimate onto the tables on Tile Sheet 15.

- Take a handful of Color Tiles. Start at one end of the room and place several tiles on the floor as if measuring the length of the room with tiles. Ask students if they can think of a better way to measure. Discuss the use of Jaguar-Strips.

- Ask a group to measure the length of the classroom using Jaguar-Tens. They should lay strips down, end-to-end, across the room. Encourage students to alternate colors to make counting by tens easier. Children should use Jaguar-Ones when they no longer can use their Jaguar-Tens. Write down the measurement to the nearest tile on Tile Sheet 15 Transparency.

Pictorial Activities

- Ask children to draw a picture of the activity on their Journal Page. They should include the number of strips and the colors.

- Groups use Jaguar-Tens and Jaguar-Ones to measure the objects on Tile Sheet 15. Two free spaces are at the bottom.

Symbolic Activities

- After each group completes their table, discuss and compare measurements. Are all the measurements the same? If not, why? (Possible answers: error in measuring, counting, differences in strips, points measured, etc.)

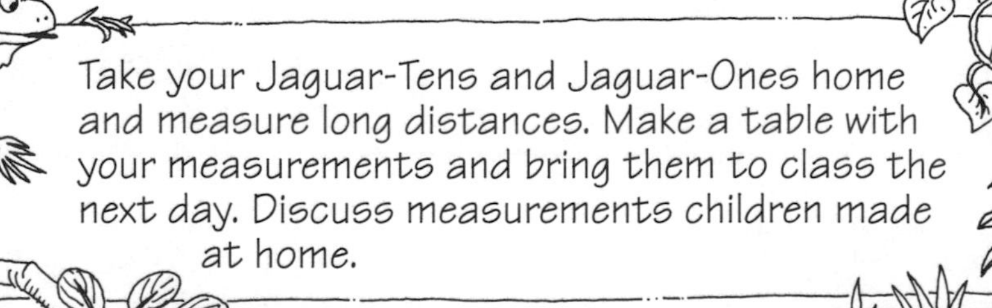

Name ____________________

TILE SHEET 15

1. Estimate the length of each object listed.
2. Now use your Jaguar-Ones and Jaguar-Tens to find the actual length of each object. Measure to the nearest tile.
3. Compare the estimates with the actual lengths.

Object	Estimated Length	Actual Length
Classroom	tiles	tiles
Student desk	tiles	tiles
Teacher's desk	tiles	tiles
Blackboard	tiles	tiles
Hall	tiles	tiles
	tiles	tiles
	tiles	tiles

AREAS OF EMPHASIS: Area, Estimating, Measurement, Counting, and Conservation of Area

Estima-Tiles

TEACHER MATERIALS

- Color Tiles
- Tile Sheet 16 Transparency
- Transparency 3
- Color Tile Grid Transparency

STUDENT MATERIALS

- Color Tiles (divided equally among groups)
- Tile Sheet 16
- Colored pencils or crayons
- Color Tile Grid
- Hundreds mat (optional - 2 pages)
- Journal Page

Concrete Activities

- What is area? Discuss area with children asking them to define it. Explain: "When we talk about area in math, we often mean the region covered by something. If I place this piece of paper on the desk, it covers a certain amount of area. If I place.these objects.

- On Transparency 3, point to the rectangle in Figure A. Ask students to estimate how many tiles will fit in this rectangle. Write down the estimate. Now try it. How many tiles make up this rectangle? (Answer: 2 square tiles) Write down the area. Repeat for Figure B (4 square tiles) and Figure C (11 square tiles).

- Ask children to draw three shapes on the Color Tile Grid, one with an area of 6 square tiles, another with an area of 8 square tiles, and a third with an area of 10 square tiles. Children should label each figure with the area. Discuss why some shapes look different but have the same area.

Pictorial Activities

- Children complete Tile Sheet 16 by matching each shape with its area. Point out that it is not necessary to use tiles to measure. Two answers are impossible. (1 and 200) The smallest shape has the smallest area, and the largest shape has the largest area.

Symbolic Activities

How many tiles fit on a hundreds-square? (100) Is the area of your desk less than, equal to, or more than the area of a hundreds-square? Use your hundreds-square to measure other large areas. Why is it useful to have a hundreds-square?

Name ____________________

TILE SHEET 16

Draw a line from each of the shapes on the left to its matching area on the right. Hint: You won't use all of the areas on the right.

1 square tile

2 square tiles

5 square tiles

12 square tiles

200 square tiles

GEOMETRY

AREAS OF EMPHASIS: Area, Estimating, Measurement, and Counting

Turtles and Tiles

TEACHER MATERIALS
- Color Tiles
- Tile Sheet 17 Transparency
- Transparency 4

STUDENT MATERIALS
- Color Tiles (divided equally among groups)
- Tile Sheet 17
- Colored pencils or crayons
- Journal Page

Concrete Activities

- Use the top half of Transparency 4. Ask: "How can I find out how many squares fit inside the turtle's shell?" Ask a volunteer to fit Color Tiles onto the turtle's shell on the overhead, and report how many were used.

- Ask: "Are turtle shells really shaped like rectangles? (No) Here is a drawing of what a real shell might look like." Show bottom half of Transparency 4. Ask: "How many tiles would it take to completely cover the turtle shell?" Ask for estimates. Have a volunteer cover the turtle shell completely with Color Tiles, and estimate the area of the second turtle shell. Ask children to think and talk about areas of other irregular shapes.

Pictorial Activities

- Divide children into groups. Ask groups to trace one member's foot on a piece of paper. Ask: "How many tiles will it take to cover the foot?" Enter children's estimates into the column labeled *Estimate (no tiles).*

- Groups trace one of their member's hands and estimate the number of tiles they think it will take to cover the hand. Enter the estimates on the table. Do the same with 3 unusually shaped objects, such as an odd paper weight or precut irregular shapes.

- Groups now use Color Tiles to find new estimates for each of the objects. Have groups enter their estimates in the last column.

Symbolic Activities

- Write down the estimates for each foot on the overhead. Record guesses about the area of all feet in the classroom. Ask: "How can we find the area of all the feet? (Answer: by adding)" Encourage the use of calculators for this problem.

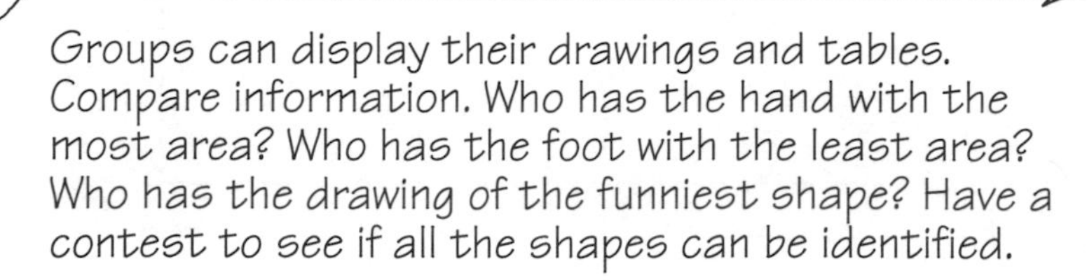

Groups can display their drawings and tables. Compare information. Who has the hand with the most area? Who has the foot with the least area? Who has the drawing of the funniest shape? Have a contest to see if all the shapes can be identified.

Name ____________________

TILE SHEET 17

Object	Estimate (no tiles)	Estimate (with tiles)
Foot	tiles	tiles
Hand	tiles	tiles
	tiles	tiles
	tiles	tiles
	tiles	tiles

GEOMETRY 3 Shape-Ominoes

TEACHER MATERIALS
- Color Tiles
- Grid Sheet Transparency

STUDENT MATERIALS
- Color Tiles (100 per group; give each group 1 color)
- Grid Sheet
- Colored pencils or crayons
- Journal Page

Concrete Activities

- State: "Today we will be putting squares together to make shapes. Before getting started, there is one rule for today: When you put squares together, place them edge-to-edge."

OK Not OK Not OK

- Begin with 1 tile. "How many different shapes can you make with 1 tile?" (1) Tilt the square 45°.

"Is this a different shape?" (No)

- Use 2 same-colored tiles and the Grid Sheet Transparency. Ask: "How can 2 tiles be put together?" Draw 2 tiles put together on the transparency. "Can anyone think of another shape?" Some children may come up with this shape:

  "Why are these two shapes the same?" (Answers: If you rotate one of the shapes, it will match the other shape exactly; you can place one shape on top of the other and they match; they are congruent.) "A shape made from 2 squares put together is called a domino. Does anyone know what a domino is? (Show dominoes) Does the shape change as I rotate the domino?" (No)

- Use 3 same-color tiles and the Grid Sheet Transparency. State: "3 squares put together is called a tromino. How can I build a shape using 3 tiles put together?" Document children's suggestions on the Grid Sheet Transparency using Color Tiles. Outline, then shade in the shapes. Children may come up with the following shapes:

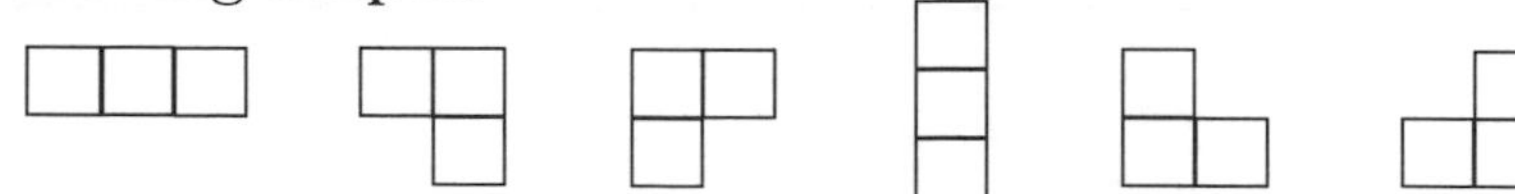

  Discuss the differences and similarities of the shapes. Some shapes are easily recognized as being the same, while others may be more difficult.

- With 3 tiles put together, emphasize that if you can rotate one shape and it matches another, then they are the same. There are only 2 possible shapes that can be created using 3 tiles put together:

- Have children work in groups. State: "4 squares put together are called tetrominoes. Build as many different shapes as possible using 4 squares put together."

Pictorial Activities

- Ask students to draw the shapes they built using 4 squares put together (on a copy of the Grid Sheet). Ask for the area of each shape. The primary purpose of this lesson is to demonstrate that the area is the same for all of the 5 possible shapes:

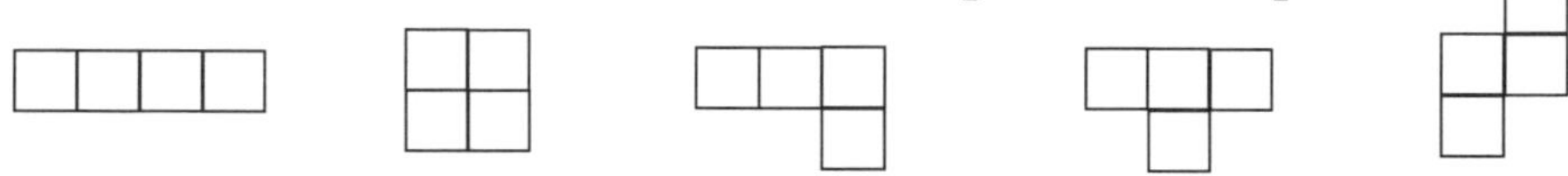

- If you can cut out one of the shapes and rotate it or flip it to cover another shape, then the shapes are the same. You can cut out the shaded tetrominoes on the Grid Sheet Transparency and rotate and flip to prove that shapes are the same (or congruent).

Concrete Activities

- 5 squares put together are called pentominoes. Build as many different shapes as you can using 5 tiles put together.

Pictorial Activities

- Draw the shapes you built using 5 squares put together (on a copy of the Grid Sheet). Emphasize that area remains constant, and discuss the total number of shapes. For pentominoes there are a total of 12 different shapes:

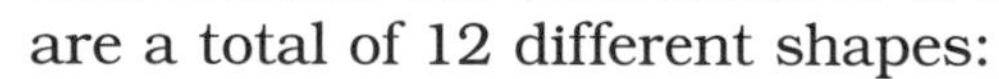

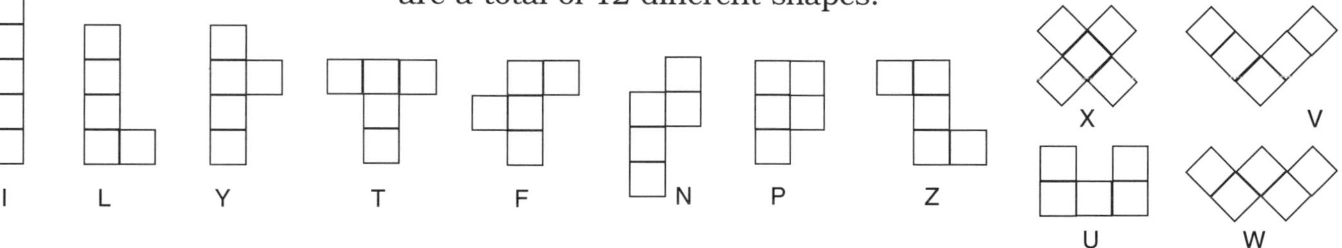

Symbolic Activities

- "Look at the shapes you have drawn with 5 squares put together. Give each of the shapes a name depending on what letter of the alphabet it looks like. Pentomino pieces are some times named according to the letter of the alphabet they resemble most. See if your class can come up with the names as shown above (or maybe even better ones).

GRAPHING AND PROBABILITY

AREAS OF EMPHASIS: Graphing, Probability, and Data Analysis

Tile Graph

TEACHER MATERIALS
- *Color Tiles*
- *Tile Sheet 18 Transparency*
- *Tile Grid Transparency*

STUDENT MATERIALS
- *Color Tiles*
- *Tile Sheet 18*
- *Journal Page*

Concrete Activities

- Take a handful of Color Tiles and place them on the overhead. State: "I am going to graph these Color Tiles." Take the first tile, show it to the class, then place it on the appropriate column. Note: Color in the box at the bottom of each row to emphasize which Color Tile you are graphing. Repeat until you have placed all tiles on the graph.

- Discuss the graph: "Which tiles do I have more of? Do I have to count to know this? (No; the tallest column has the most tiles.) Do I have to count to know which color I have the least of? (No; the shortest column has the least tiles.) Do I have the same of any two colors? (Answers will vary; yes, if two columns are of the same height.) How many tiles do I have altogether? (Count)"

- Children work in groups and use Tile Sheet 18. Have two people from each group take a handful of tiles. Groups use Tile Sheet 18 to graph the two handfuls of tiles.

Pictorial Activities

- Ask children to remove the tiles and color in the squares on each column with the corresponding colors (blue, yellow, green, red).

Symbolic Activities

- Ask children to answer the following questions on another piece of paper:

  Of which color or colors of tiles did you have most?
  Of which color or colors of tiles did you have least?
  Did you have the same amount of any two colors?
  If yes, which colors?
  How many tiles did you have altogether?

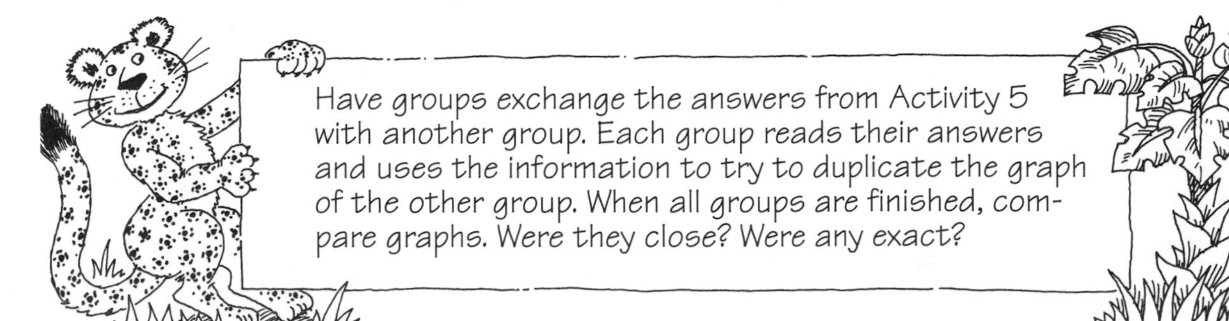

Name ____________________

TILE SHEET 18

Yellow	Blue	Red	Green

Number of Tiles

Color

GRAPHING AND PROBABILITY

AREAS OF EMPHASIS:
Graphing and Probability

Mystery Bags

TEACHER MATERIALS
- Demonstration Bag
- Tile Sheet 19 Transparency

STUDENT MATERIALS
- Group Bags
- Tile Sheet 19
- Colored pencils or crayons
- Journal Page

Concrete Activities

- Before class, prepare labeled paper bags with tiles in all 4 colors (1 bag per group). Only one of the bags should have more yellow tiles than red, blue, or green tiles. For example: 1 blue tile, 5 red tiles, 10 green tiles and 20 yellow tiles. The rest of the bags should have more of one color (blue, red, or green) than the other three colors. Your demonstration bag should have 5 red tiles, 10 blue tiles, and 15 yellow tiles.

- Show children the demonstration bag. State: "In this bag I have put red tiles, blue tiles, and yellow tiles. Is there a chance that I might pull out a blue tile? (Yes) A red tile? (Yes) A yellow tile? (Yes) A green tile? (No; there are no green tiles in the bag.)"

- Pull out a tile and record it on the Tile Sheet 19 Transparency. Place the tile back in the bag. Ask for a prediction of what the next tile will be. Choose another tile, record it, and return it. Repeat until you have at least 10 samples.

- Investigate and discuss the graph you have created. Use the graph to form predictions about what is in the bag. (There are probably more yellow tiles; there are blue, green, and yellow tiles; etc.) Take the tiles out of the bag to check predictions.

Pictorial Activities

- Use Tile Sheet 19. Pass out mystery bags. Each group should have 2 teams. One team guesses and graphs, while the other pulls tiles from the bag. Remind children to write down the bag number at the top of Tile Sheet 19. After 10 samples have been picked, the guessing team writes statements describing what is in the bag. Groups check the contents of their bags, comparing them with their graphs and sentences.

Symbolic Activities

- Ask each group to find another team with the same bag number at the top of their tile sheet. Have groups compare and discuss their graphs. Are they similar? Are they different?

Display graphs on a bulletin board. Can the graphs determine which bag has more of a certain color? Discuss with children.

Bag Number ____________ Name ________________

TILE SHEET 19

Number of Tiles

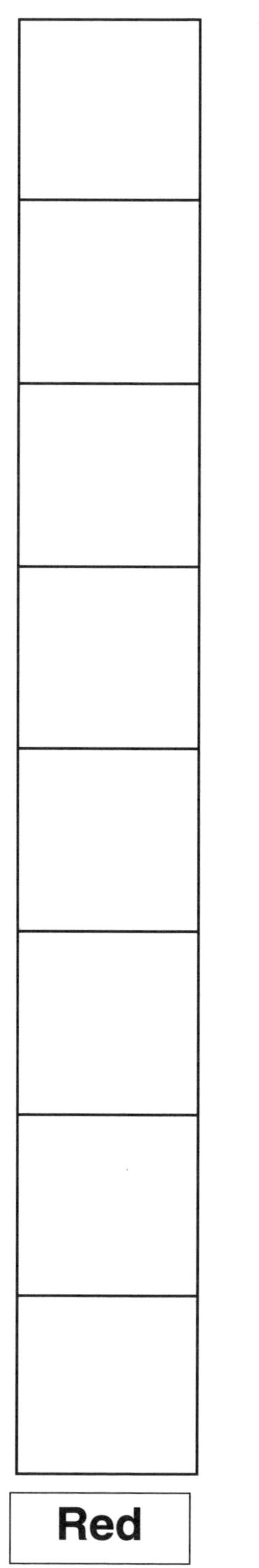

Yellow	Blue	Red	Green

Color

GAMES AND PUZZLES

AREAS OF EMPHASIS:
Logic and Patterning

Games

Game 1: Guessing Block

Materials

- *Color Tile Grid (cut in half so each player or team has a 5 x 5 gameboard)*
- *Color Tiles (at least 100, with equally selected colors)*
- *Paper Bag*

Instructions

- Each player has a blank 5 x 5 gameboard.
- Ask students to color each square either blue, red, yellow or green. They can make any design they wish.
- Place the Color Tiles into the paper bag.
- Have each student play against another.
- Each player chooses a tile without looking.
- If a player chooses a tile the same color as an uncovered space, he/she can place the tile on it. If not, he/she places the tile back in the bag.
- The person or team that covers the gameboard first wins.

Game 2: Pattern Block

Materials

- *Color Tile Grid (cut in half so each player or team has a 5 x 5 gameboard)*
- *Color Tiles (at least 100, with equally selected colors)*
- *Paper Bag*

Instructions

- Place the Color Tiles into the paper bag.
- Have each player choose a tile without looking.
- A player can place the tile anywhere on his/her gameboard as long as it is not next to another tile of the same color (vertically, horizontally, or diagonally). If a player cannot use the tile, she places it back in the bag.
- The person or team that covers the gameboard first or has the least number of blank squares left wins.

Teacher Notes: The winning strategy is to place the tiles in a pattern. You can tell students there is a hint in the title.

B	R	B	R	B
G	Y	G	Y	G
B	R	B	R	B
G	Y	G	Y	G
B	R	B	R	B

Answer to Puzzle 1 (page 47)

Y	Y	R	R
B	Y	G	R
B	B	G	G

Answer to Puzzle 2 (page 48)

B	B	B	B
Y	Y	Y	R
Y	G	R	R
G	G	G	R

PUZZLE 1

Fill the Rectangle

Use Color Tiles in these shapes to fill the rectangle below:

B B
B

Y Y
Y

R R
R

G
G G

PUZZLE 2 Fill The Square

Use Color Tiles in these shapes to fill the rectangle below:

B	B	B	B

R	R	R
	R	

G	G	G
	G	

Date ____________________ Name ____________________

JOURNAL PAGE

QUILT

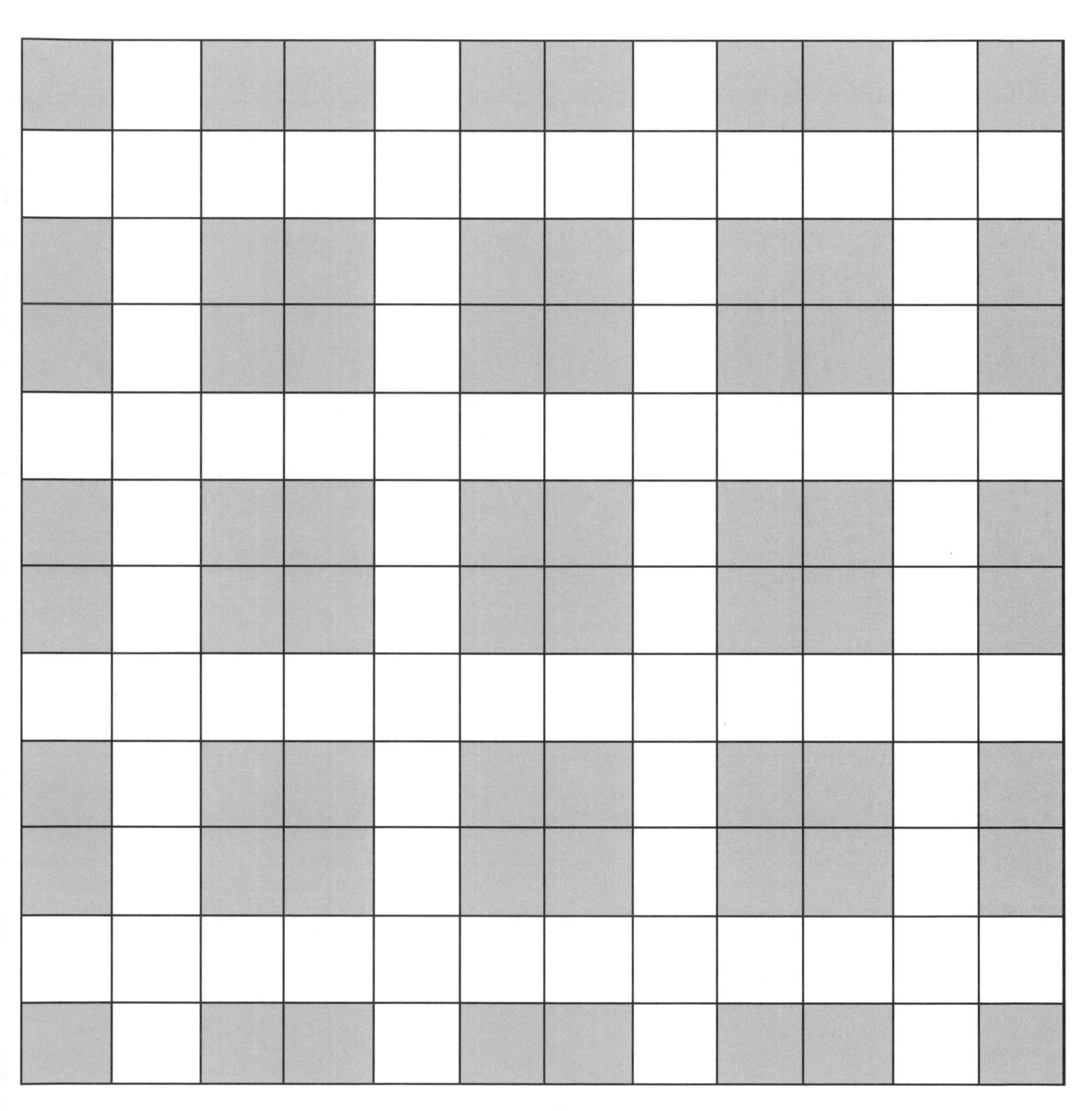

OVERHEAD COLOR SQUARE GRID

JAGUAR-ONES

JAGUAR-TENS

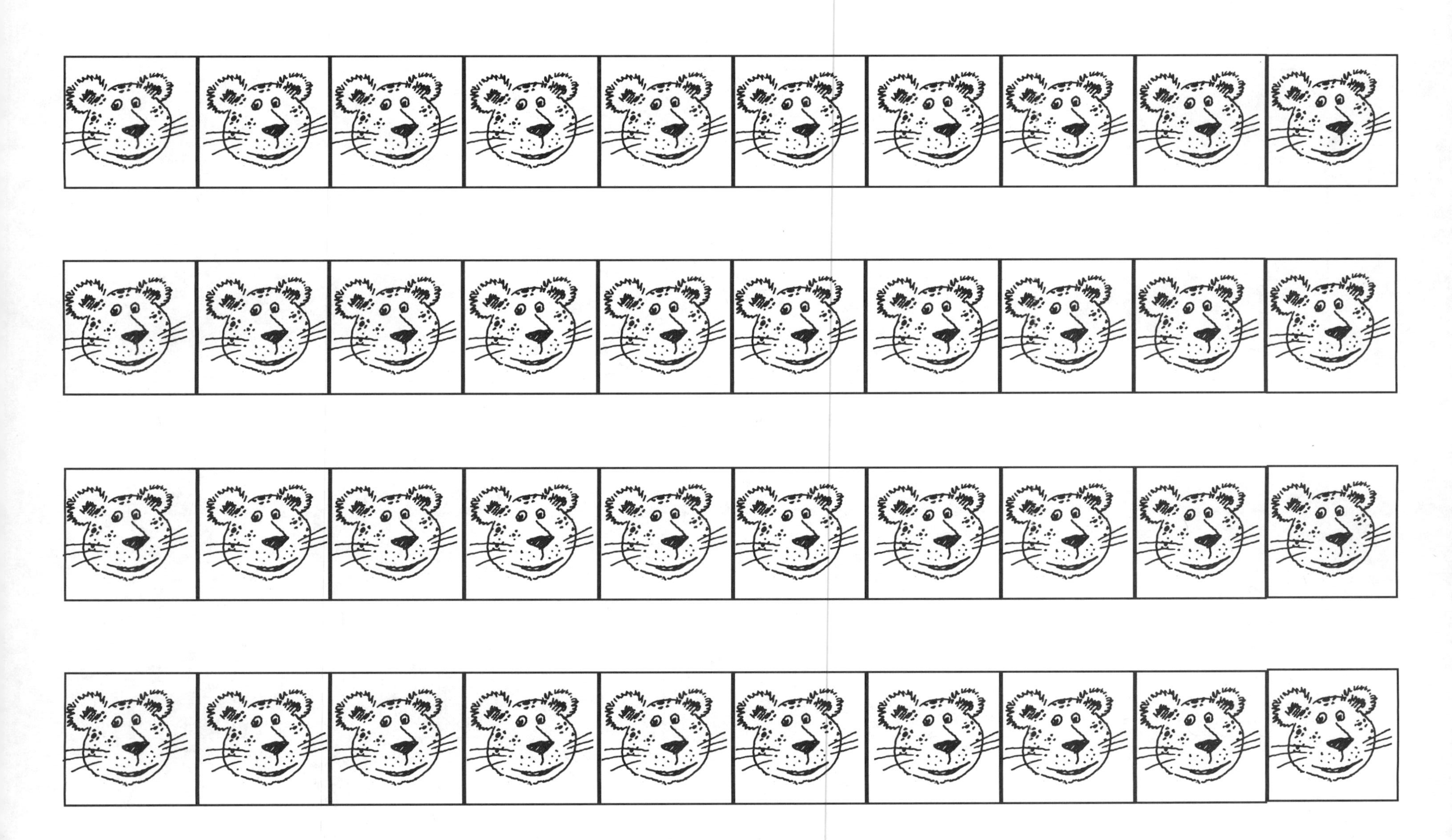

COLOR TILE GRID / HUNDREDS MAT (PART 1)

COLOR TILE GRID / HUNDREDS MAT (PART 2)

Note: To make a Hundreds Mat, tape this page to the previous page, matching the tile borders.

TRANSPARENCY 1

TRANSPARENCY 2

How many R?	How many B?	How many altogether?

__________ + __________ = __________

__________ + __________ = __________

__________ + __________ = __________

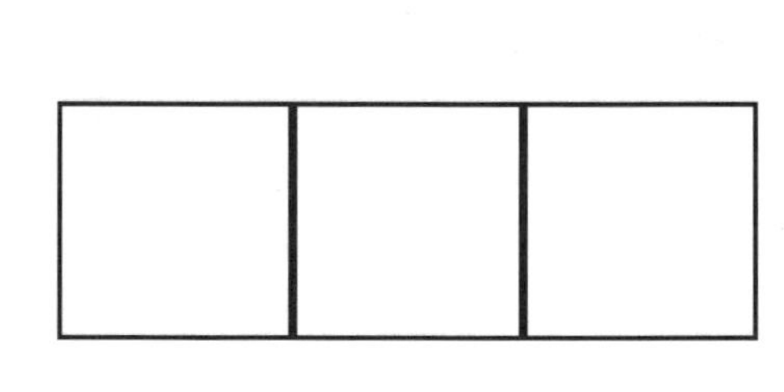

How many R?	How many B?	How many altogether?

__________ + __________ = __________

__________ + __________ = __________

__________ + __________ = __________

__________ + __________ = __________

TRANSPARENCY 3

Figure A

Estimate ______ tiles

Area ______ tiles

Figure B

Estimate ______ tiles

Area ______ tiles

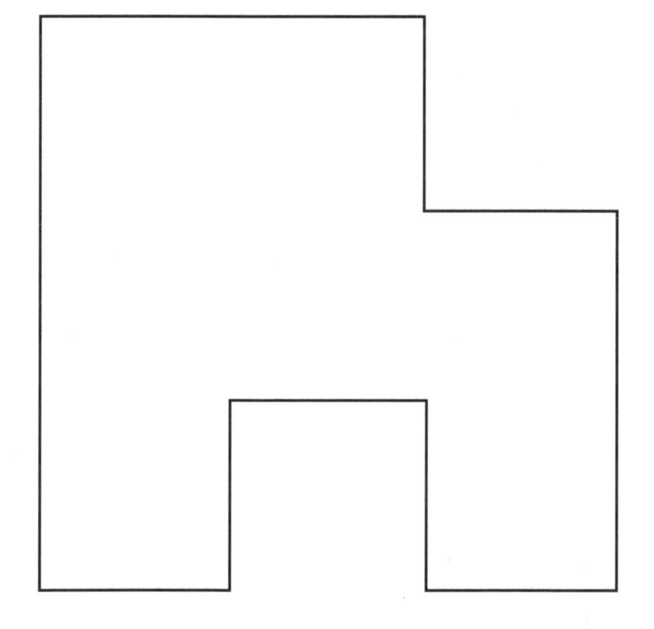

Figure C

Estimate ______ tiles

Area ______ tiles

TRANSPARENCY 4

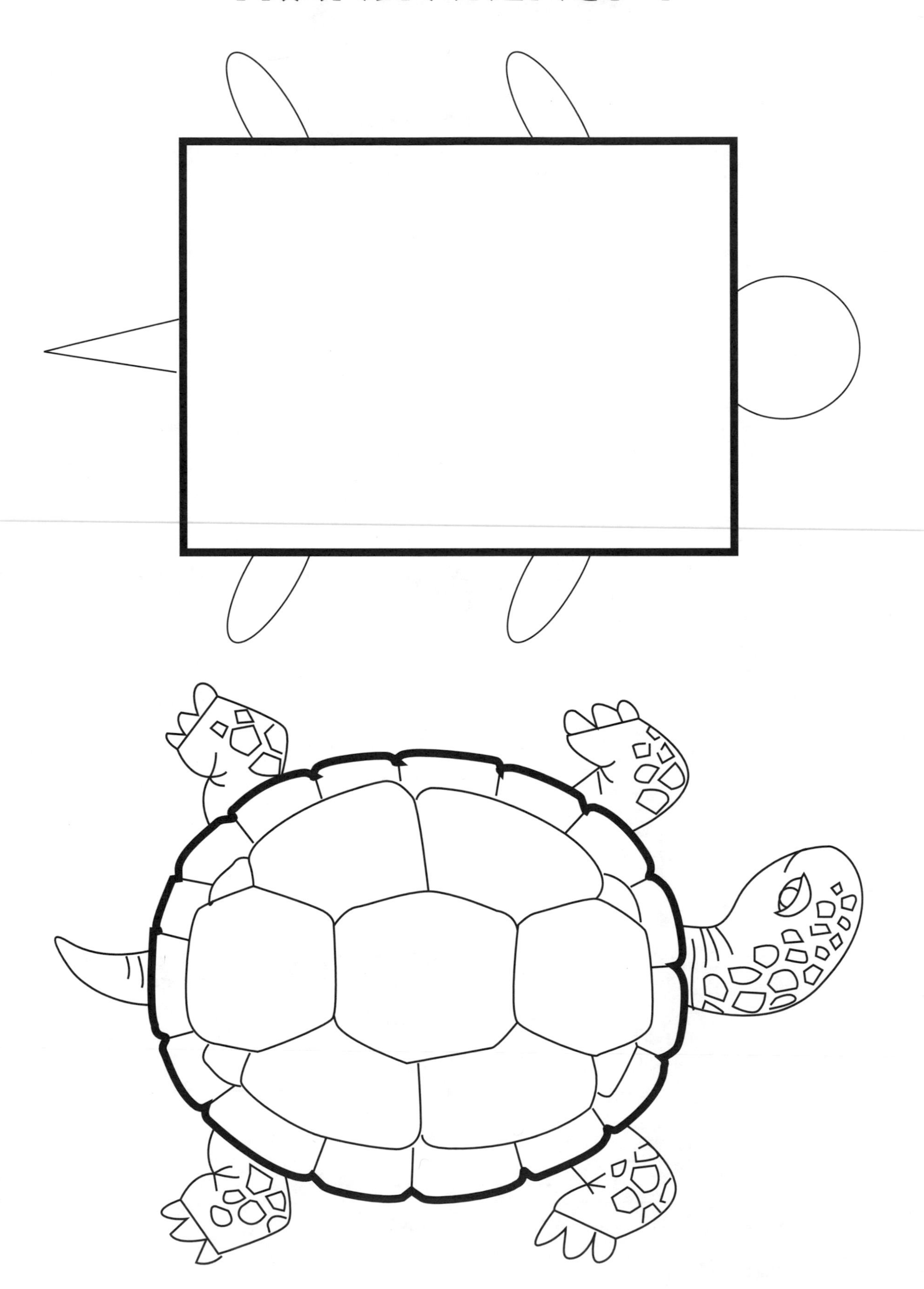

TILE SHEET 1

B	Y	B	Y	B	Y	B	Y

Y	B	Y	B	Y	B	Y	B

B	Y	B	Y	B	Y	B	Y

Y	B	Y	B	Y	B	Y	B

Name ___________

TILE SHEET 2

A

B	Y	B	Y	B	Y	B	Y
Y	B	Y	B	Y	B	Y	B

B

B	Y	B	Y	B	Y	B	Y
Y	B	Y	B	Y	B	Y	B

C

B	Y	B	Y	B	Y	B	Y
Y	B	Y	B	Y	B	Y	B

Name ___________

TILE SHEET 3

Copy and continue the pattern.

B	R	Y	B
R	Y	B	R
Y	B	R	Y
B	R	Y	B

Create your own pattern.

Name ___________

TILE SHEET 4

Draw the tiles that belong in each bag (from 1 to 5).
Write the number of tiles in each bag on the lines below.

1 ___ 2 ___ 3 ___ 4 ___ 5 ___

Name ____________

TILE SHEET 6

1. Color the squares yellow to show each number.

5	
9	
15	
18	

2. Give the numbers that stand for the tiles shown.

	13
	5
	18
	8
	23

Name ____________

TILE SHEET 7

How many tiles are in each bag? ______

1. Draw the bags and tiles that stand for the numbers in the following table. The first one is finished.

Number	Bags	Tiles
15		
19		
22		
32		
34		
47		

2. Answer the following questions.

Sarah has 1 bag and 8 tiles.
How many tiles does Sarah have? **18**

Anthony has 2 bags and 7 tiles.
How many tiles does Anthony have? **27**

Taylor has 3 bags and 5 tiles.
How many tiles does Taylor have? **35**

Kyle has 4 bags and 6 tiles.
How many tiles does Kyle have? **46**

Name ____________

TILE SHEET 8

Yellow = 1 Blue = 10

1. Color the squares blue or yellow to show each number.

18 B Y Y Y Y Y Y Y Y
24 B B Y Y Y Y
39 B B B Y Y Y Y Y Y Y Y Y
41 B B B B Y
52 B B B B B Y Y

2. Draw a line from each tile combination to its matching number.

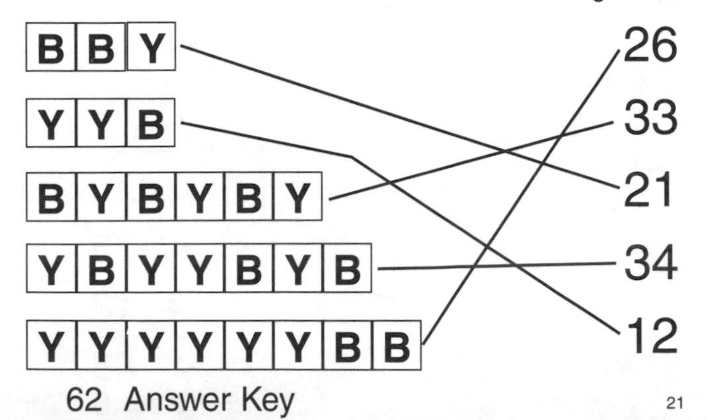

Name ____________

TILE SHEET 9

1.

Color of Tile	Draw the Tiles	How many?
Yellow		10
Blue		6
Green		4
Red		2

2. Add the tiles.

Y and B **16**
Y and G **14**
Y and R **12**

B and Y **16**
B and G **10**
B and R **8**

G and Y **14**
G and B **10**
G and R **6**

R and Y **12**
R and B **8**
R and G **6**

Name ____________

TILE SHEET 10

Use red and blue tiles. Find the different combinations.

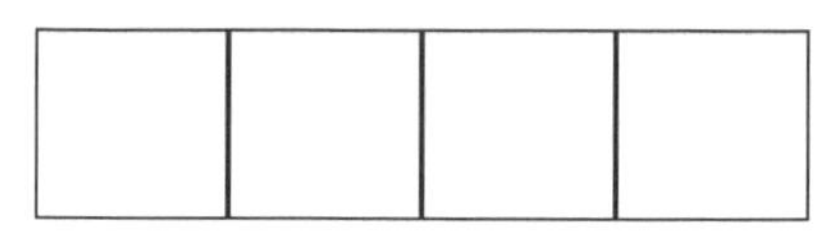

Draw your combinations here.

	How many R?	How many B?	How many altogether?
RRRR	+	=	
RRRB	+	=	
RRBB	+	=	
RBBB	+	=	
BBBB	+	=	

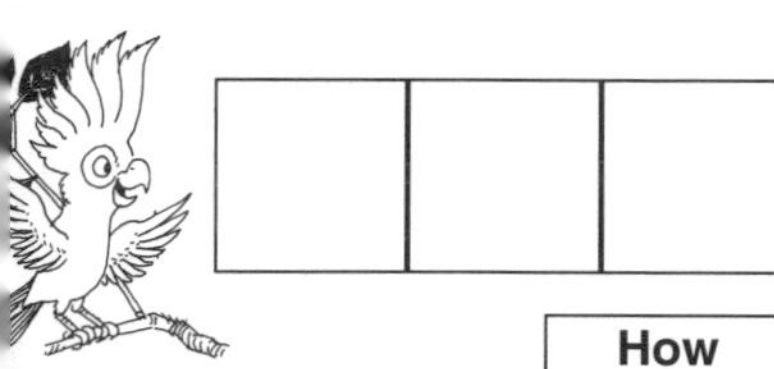

Draw your combinations here.

	How many R?	How many B?	How many altogether?
RRRRR	+	=	
RRRRB	+	=	
RRRBB	+	=	
RRBBB	+	=	
RBBBB	+	=	
BBBBB	+	=	

Name ____________

TILE SHEET 11

Draw in the missing tiles.

1.
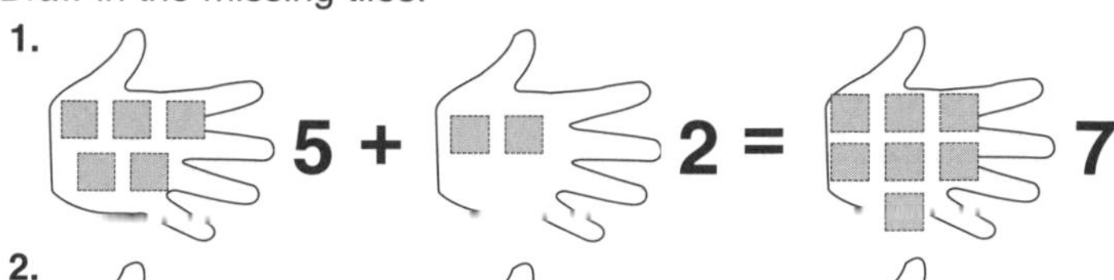
5 + 2 = 7

2.
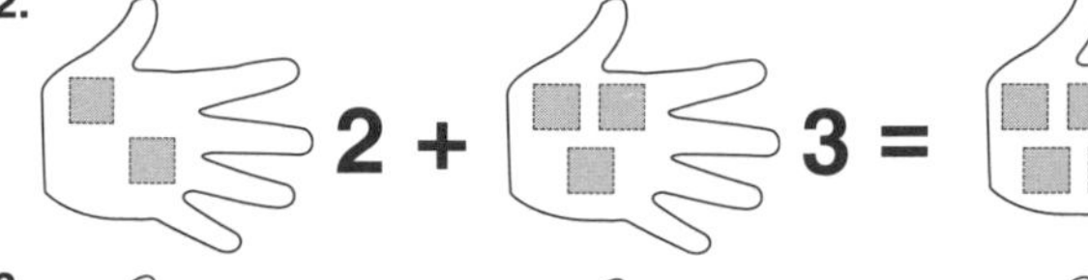
2 + 3 = 5

3.
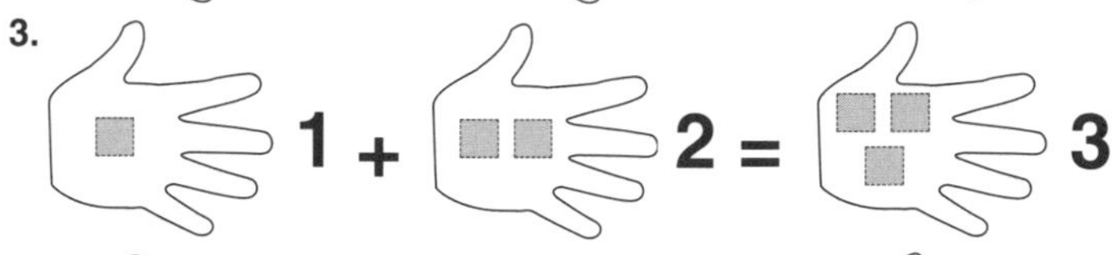
1 + 2 = 3

4.
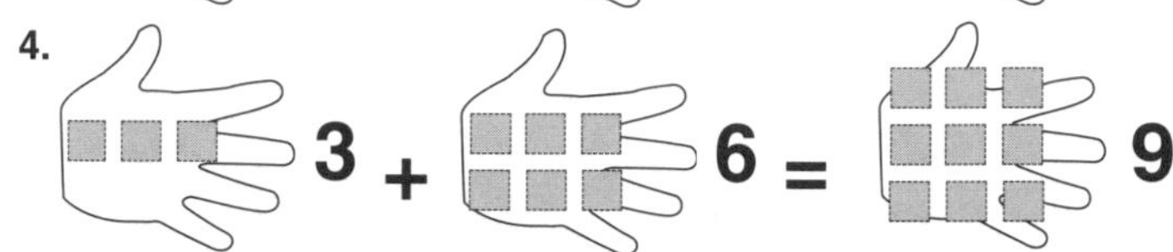
3 + 6 = 9

5.
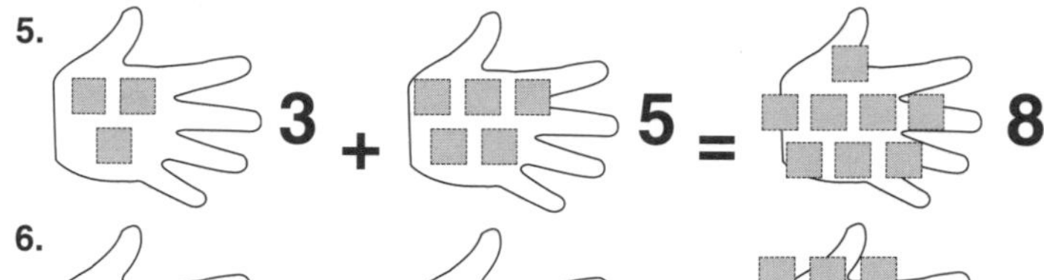
3 + 5 = 8

6.
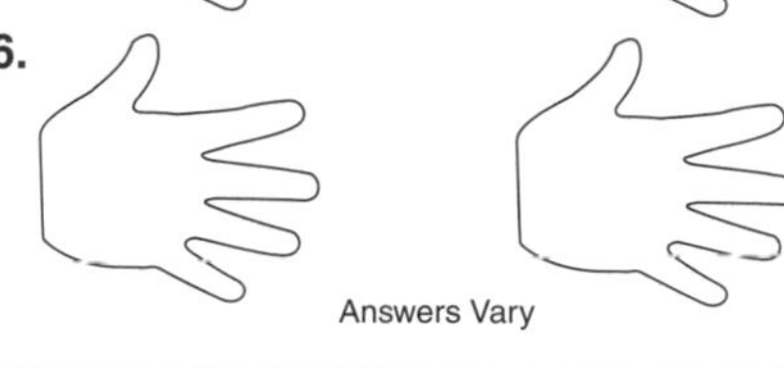
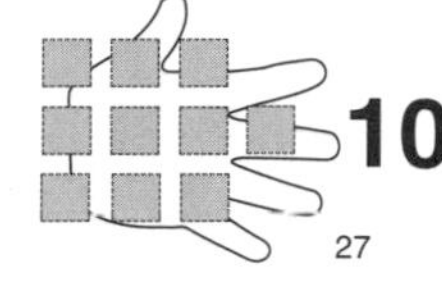
10

Answers Vary

Name ____________

TILE SHEET 12

1. Draw the tiles to show the different ways you can group 6. The first two are done for you.

Group tiles like this	First group	Second group
6 and 0	□□□□□□	
5 and 1	□□□□□	□
4 and 2	□□□□	□□
3 and 3	□□□	□□□
2 and 4	□□	□□□□
1 and 5	□	□□□□□
0 and 6		□□□□□□

2. Complete the following problems.

6 - 0 = 6

6 - 1 = 5

6 - 2 = 4

6 - 3 = 3

6 - 4 = 2

6 - 5 = 1

6 - 6 = 0

Name ____________

TILE SHEET 13

Yellow = 1 Blue = 10

1. Add the blue and yellow tiles below. Draw the tiles to show the answer. Then write the answer to the problem.

Example:

BYYYY 14
BYYYY + 14
28
BB YYYYYYYY

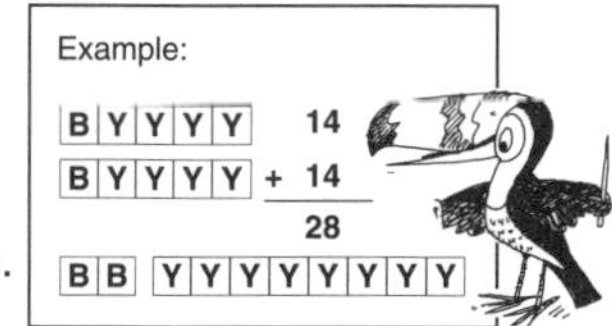

a)
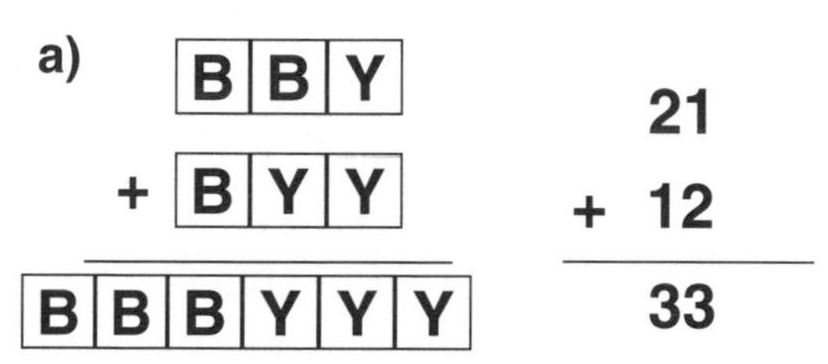
BBY + BYY = BBBYYY

21 + 12 = 33

b)
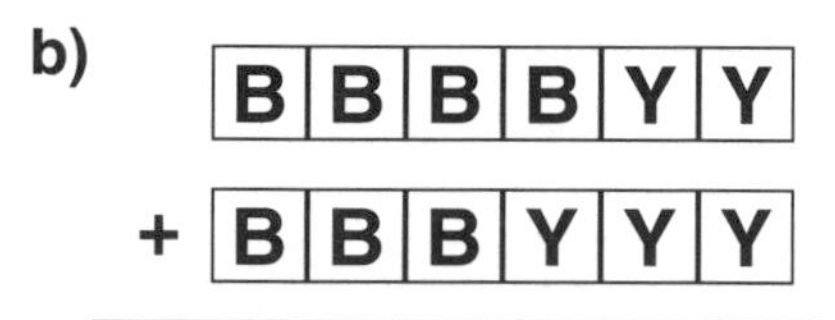
BBBBYY + BBBYYY = BBBBBBBYYYYY

42 + 33 = 75

2. Solve each of the following problems.

22 + 63 = 85

38 + 51 = 89

75 + 13 = 88

Name ______________

TILE SHEET 16

Draw a line from each of the shapes on the left to its matching area on the right. Hint: You won't use all of the areas on the right.

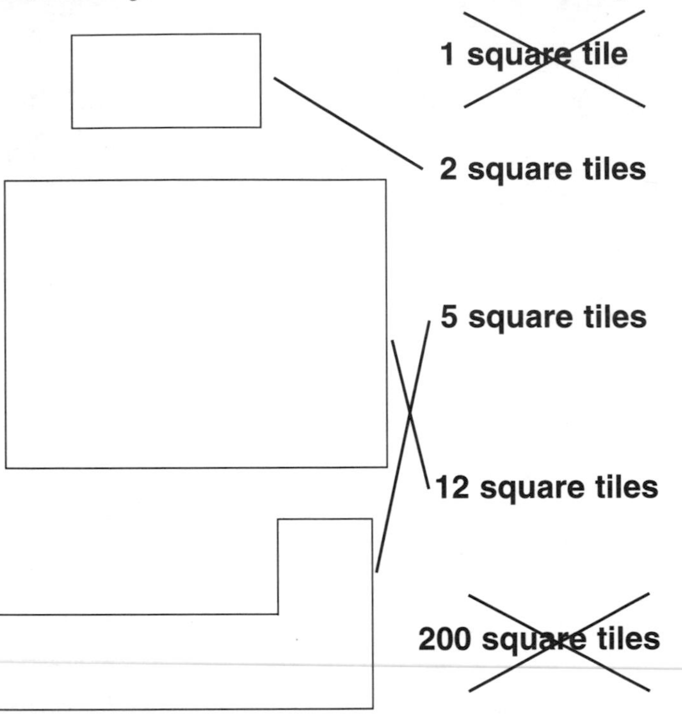

37

TRANSPARENCY 2

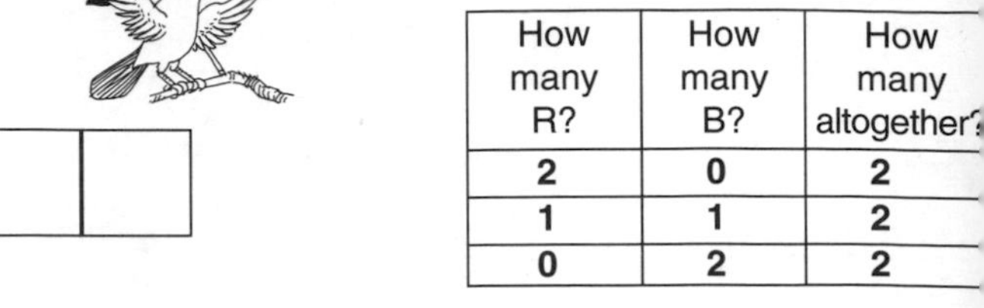

How many R?	How many B?	How many altogether?
2	0	2
1	1	2
0	2	2

2 + 0 = 2
1 + 1 = 2
0 + 2 = 2

How many R?	How many B?	How many altogether?
3	0	3
2	1	3
1	2	3
0	3	3

3 + 0 = 3
2 + 1 = 3
1 + 2 = 3
0 + 3 = 3

58

TRANSPARENCY 3

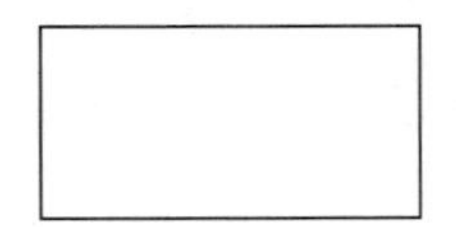

Figure A

Estimate _____ tiles

Area 2 tiles

Figure B

Estimate _____ tiles

Area 4 tiles

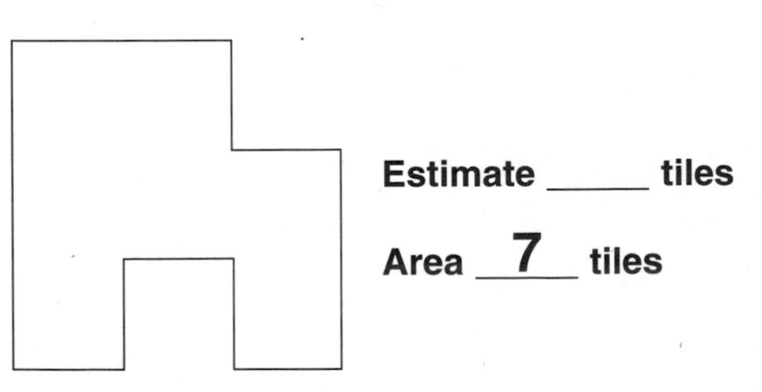

Figure C

Estimate _____ tiles

Area 7 tiles

59

TRANSPARENCY 4

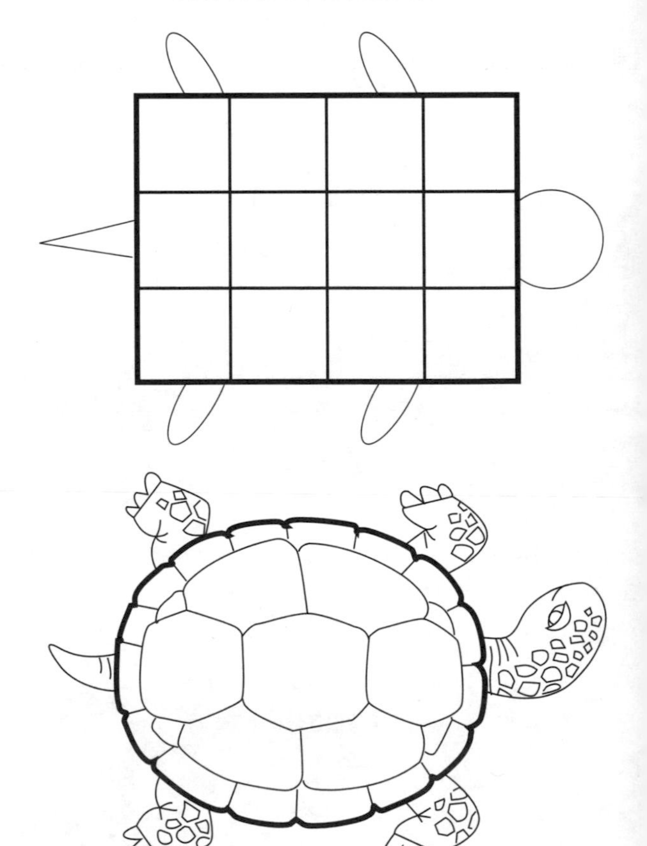

60